APPROPRIATE PRODUCTS, EMPLOYMENT AND TECHNOLOGY

Low-income consumers in developing countries do not derive full benefit from their purchases. This is so because most new products embody characteristics corresponding to consumer preferences in rich rather than poor countries. In addition, strong advertising for these products and the relatively low educational level of the low-income consumer may further aggravate this situation. This book, which contains eight new case-studies on basic needs goods (including soap, furniture, weaning foods, bicycles, footwear, passenger transport and household utensils – mainly in Asia and Africa) demonstrates further that low-income consumers usually spend a large part of their income on products and services produced by small-scale enterprises, thereby generating a considerable amount of employment. There is also strong evidence that small enterprises are normally not able to reach the high-income consumer market. By contrast, large enterprises do have many low-income customers for products such as footwear, soap and weaning foods. Thus one of the important policy implications of this work is that more sustained support to small-scale and rural industries might itself lead to an increased demand for relatively labour-intensive products. Other suggested policies, such as consumer education, appropriate product standards and trade marks, and objective product information, aim at increasing the efficient consumption of low-income classes.

Dr Wouter van Ginneken is Senior Economist, Bureau of Labour Problems Analysis, ILO, Geneva, and the author of *Rural and Urban Income Inequalities in Indonesia, Mexico, Pakistan, Tanzania and Tunisia* and *Socio-economic Groups and Income Distribution in Mexico*.

Mr Christopher Baron is Senior Programme Analyst, Bureau of Programming and Management, ILO, Geneva, and the editor of *Technology, Employment and Basic Needs in Food Processing in Developing Countries*.

The World Employment Programme (WEP) was launched by the International Labour Organisation in 1969, as the ILO's main contribution to the International Development Strategy for the Second United Nations Development Decade.

The means of action adopted by the WEP have included the following:

- short-term high-level advisory missions;
- longer-term national or regional employment teams; and
- a wide-ranging research programme.

Through these activities the ILO has been able to help national decision-makers to reshape their policies and plans with the aim of eradicating mass poverty and unemployment.

A landmark in the development of the WEP was the World Employment Conference of 1976, which proclaimed, *inter alia*, that "strategies and national development plans should include as a priority objective the promotion of employment and the satisfaction of the basic needs of each country's population." The Declaration of Principles and Programme of Action adopted by the Conference have become the cornerstone of WEP technical assistance and research activities during the 1980s.

This publication is the outcome of a WEP project.

APPROPRIATE PRODUCTS, EMPLOYMENT AND TECHNOLOGY

Case-Studies on Consumer Choice and Basic Needs in Developing Countries

Edited by

Wouter van Ginneken

and

Christopher Baron

A study prepared for the International Labour Office within the framework of the World Employment Programme

MACMILLAN PRESS
LONDON

First published 1984 by
THE MACMILLAN PRESS LTD
London and Basingstoke
Companies and representatives
throughout the world

ISBN 0 333 35302 1

Printed in Hong Kong

Contents

Preface

This volume is the result of an extensive research project launched shortly after the Declaration of the ILO's World Employment Conference in 1976. The central objective of this project was to determine the extent to which an improvement in income distribution (by various means, including policy measures directly aimed at the alleviation of poverty in the rural areas of developing countries) is likely to be consistent with the consumption of goods and services produced by techniques that are relatively labour-intensive and therefore employment-creating. Such consistency between the consumption and production sides of the economy has been a key assumption in the work of many of the ILO's employment strategy missions to developing countries, though it has never been rigorously tested at the product level. Most previous studies of production in developing countries have focused on technology and its appropriateness in these countries. In contrast, an important and distinctive contribution of this collection of case-studies is that each case-study also considers consumer behaviour.

The conclusion to be drawn from these studies, viewed together, is that if the most appropriate products and services are available, low-income consumers satisfy their basic needs in an efficient manner. However, appropriate products are not always available, either because they are not produced in a certain country, or because they are not available in quantities, or at prices, that a low-income consumer would judge appropriate. Another important finding is that low-income consumers usually spend a large part of their income on products and services produced by small-scale enterprises, thereby generating a considerable amount of employment. There is further strong evidence that small-scale enterprises are normally not able to reach the high-income consumer market. However, large-scale enterprises do have many low-income consumers as their customers for products such as footwear, soap and weaning food. Thus one of the important policy implications of this work – others are spelt out in the final chapter – is that more sustained government support to small-scale and rural industries might itself lead to an increased demand for relatively labour-intensive products.

The ILO acknowledges with gratitude the research grant from the Government of Sweden (Swedish Agency for Research Cooperation with Developing Countries – SAREC) from which the country studies were financed.

AJIT S. BHALLA

Technology and Employment Branch
International Labour Office

Acknowledgements

This volume was prepared between 1977 and 1982, when both the editors were working in the ILO's World Employment Programme. We would like to thank all those who assisted its preparation. First and foremost, thanks are due to the authors of the individual chapters who invested much time in carrying out the consumer and producer surveys and then writing their research reports. They were patient in accommodating many comments and queries from the editors. Second, the many firms that were consulted during the research are gratefully acknowledged. Third, particular thanks are due to Ajit Bhalla, who, together with Felix Paukert, encouraged us throughout the project with valuable comments and advice. Fourth, Jeffrey James, Enyinna Chuta and Gijsbert van Liempt read the whole manuscript and gave further useful suggestions for its improvement. Finally, Mrs Italici, and the ATS group of the ILO typing pool, accepted and retyped – patiently and skilfully – the various drafts of the manuscript.

WOUTER VAN GINNEKEN
CHRISTOPHER BARON

1 Introduction[1]

Wouter van Ginneken and Christopher Baron[2]

In 1976 the ILO's World Employment Conference adopted the so-called 'basic-needs approach' to development. According to this approach, 'development planning should include, as an explicit goal, the satisfaction of an absolute level of basic needs'.[3] The main policies that seemed to be implied by this approach were land reform, and the application of appropriate technology in order to increase productive employment, on the one hand, and the provision of government services such as education and health, in order to improve the condition of the poorest in society, on the other. A relatively neglected implication of the basic-needs approach has been the question of product choice and appropriate products. It is to a greater knowledge of this aspect that this book is intended to contribute. Although it is the first – to our knowledge – that deals systematically with product choice in developing countries, two groups of researchers have contributed in a fundamental way to the analytical framework which is the basis of the eight case-studies included in this volume.

The work of the first group of researchers has been brought together by James and Stewart, who summarised the results of eight case-studies on new products in developing countries in a recent article.[4] The basic thesis of these authors is that most new products in the developing countries embody characteristics in proportions corresponding to consumer preferences in rich countries rather than in poor ones. In consequence, low-income consumers in developing countries do not derive maximum basic-needs satisfaction from their purchases. James and Stewart also analysed the impact of new products on technology, employment and demand for substitute products. In only one of these product studies, that concerning soap, was there clear evidence of a new product actually reducing the demand for established products. In at least six out of the eight cases, promotional expenditure was probably responsible for an over-valuation by consumers of the benefits of the new products leading to over-consumption. In seven of the eight cases, the new products had more high-income characteristics than the established ones, with the result that their impact was inegalitarian. Plastic sandals was the only case of a new low-income product, the consumption impact of which appeared to be egalitarian. In every case, including sandals, the technology was inappropriate compared with the technology associated with the older products. The main positive effect was the extension of consumer choice and the benefits accruing to those (mainly high-income) consumers who prefer the new products to the traditional ones.

The second group[5] of research studies that inspired the present volume each examined the income distribution–consumption–technology–employment hypothesis. According to this hypothesis, a reduction in income inequality leads to an expenditure pattern favouring the consumption of goods and services produced by relatively labour-intensive technologies, thereby generating employment. Macroeconomic studies examining this hypothesis have tended to indicate that a progressive redistribution of income has a positive – albeit small – impact on employment. This can be explained partly in terms of two opposing influences. On the one hand, the increased income of low-income groups is spent on agricultural goods which are typically produced with a more labour-intensive technology than manufactured goods. On the other hand, the reduced consumption expenditure of high-income groups diminishes the total demand for labour-intensive services. Paukert and others[6] also found that indirect employment effects are much more important than the direct effects, and that the reduction of the savings ratio explains about half the calculated increase in employment, whereas the impact of the shift towards more demand for labour-intensive products accounts for only a quarter of this increase. Finally, an unexpected result is that redistribution does not necessarily lead to a reduction in imports, because food imports are likely to rise.

According to James,[7] it is not especially surprising that the employment effects estimated by the macroeconomic studies are small, because they limit their analysis to broad sectors of activity including products produced with relatively capital- and labour-intensive technologies. That is, he argues that systematic differences in labour-intensity of goods consumed by the 'rich' and 'poor' do not, in general, emerge at the aggregative levels of analysis (usually the SITC two-digit level) at which most of the research has been conducted. In his case-study on the Indian sugar-processing industry he shows that the aggregation approach – taking crystal sugar (capital-intensive) and gur (labour-intensive) together – underestimates the employment effects of changes in income distribution by more than 50 per cent compared with a disaggregative approach that considers the demand for gur and sugar separately.

The case-studies included in this book were expressly designed to test the income distribution–technology–employment hypothesis for particular products, that is, at a disaggregative level, and examine whether the products currently available on the markets of developing countries satisfy the basic needs of low-income consumers in an efficient manner. The direction of causation implied by the income distribution–technology–employment hypothesis and its relation to the satisfaction of basic needs is shown in Figure 1.1.

Schematically, one can consider basic-needs satisfaction to be determined by two factors; first of all, the availability of products which given their price efficiently satisfy basic needs (see section 1.1), and secondly, the level of income usually gained through employment (see section 1.2). Basic-needs satisfaction can also be enhanced indirectly, that is, by new products that allow the greater satisfaction of basic needs at the same or at a lower price and/or by new technologies that increase the optimal use of available resources.

The impact of marketing on consumer choice and thereby on basic-needs satisfaction is an additional issue (see section 1.3) which is examined by some of the case-studies in this volume. The analytical framework also brings out some important relationships that are not analysed by the case-studies. First of all, the link between technology and income distribution so crucial to the work of Stewart[8] is not examined. Stewart shows that the application of modern capital-intensive technology increases income inequality between the modern and traditional sectors. In her view, in the context of the prevailing socioeconomic conditions in the world, inequality tends to be self-perpetuating since products bought by high-income groups are generally produced applying a relatively capital-intensive technology. Secondly, the case-studies can only partly examine the income distribution–technology–employment hypothesis since they are not able to capture the indirect employment effects resulting from income redistribution.

1.1 Basic Needs and Appropriate Products

According to the *Declaration of Principles and Programme of Action Adopted by the World Employment Conference*[9] in 1976, basic needs

4

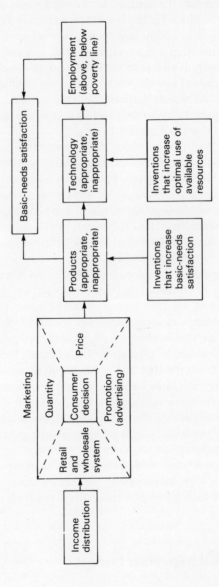

FIGURE 1.1 Analytical framework for the case-studies

include two groups of products and services:

> First, certain minimum requirements of a family for private consumption: adequate food, shelter and clothing, as well as certain household equipment and furniture. Second, essential services provided by and for the community at large, such as safe drinking water, sanitation, public transport and health, educational and cultural facilities.

Clearly, only consumer goods in the first group mentioned here (i.e. essential durables and non-durables) will be examined by the case-studies because these are purchased by individual consumers or consuming households.

Related to the concept of basic needs is the concept of appropriate products. Economic theory implicitly assumes that every product purchased is 'appropriate': if consumers are willing to pay for a product it must by definition satisfy some need of the consumer. However, not every need is obviously a basic need,[10] and many consumers in developing countries do not have an income sufficient to meet all their basic needs, however defined. Moreover, it is difficult to determine objectively the extent to which a certain product satisfies basic needs. However, the theory of consumer demand developed by Lancaster[11] may usefully be brought into the discussion in order to make the concept of appropriate products (i.e. products that efficiently satisfy basic needs) more precise.

The novelty of Lancaster's approach lies in the assumption that it is not products themselves that are objects of utility (the assumption of the traditional theory), but rather the properties or the characteristics embodied in the products. This implies that each purchaser derives his satisfaction indirectly, through the desirable characteristics of the products they consume. It is further assumed in the theory that the characteristics possessed by a good (or a combination of goods) are the same for all consumers. Thus the personal element in consumer choice (i.e. his or her utility function) is manifested only in the choice between combinations of characteristics, and not in the individual's perception of the characteristics embodied in the goods.

An illustrative example adapted to the situation in developing countries may illustrate some of the key implications of Lancaster's approach. Suppose that three people go to an eating house after a day's work (not having eaten before) and that each of them has the same amount of money to spend on a meal. The first, a day labourer, has been digging an irrigation ditch for the whole day; the second, a manual worker, has worked on an assembly line; and the last, a clerical employee, has been working in an office. The eating house offers five different meals for the same price. The difference between the meals is that some include more starches and fat, whereas others include more meat. In other words, the two main characteristics of the meals are the amounts of calories and proteins. The five different combinations of calories and proteins included in the meals are shown in Figure 1.2. Because of the different types of work in which each person is engaged, the day labourer is likely to prefer meal 2, the manual worker meal 3 and the clerk meal 4. The food requirements of the employee (2700 calories) and the manual worker (3000 calories) are met, while the protein requirements of the day labourer are not. The reason is that the needs of the day labourer (3500 calories) (or, in Lancaster's terms, his utility function) are different from those of the manual worker and the clerk.

In this example the consumer can only choose between five discrete possibilities. For other goods that have common characteristics (i.e. substitutes for many purposes) the consumer may have more choice. For example, a housewife may be able to buy rice, maize or beans (which belong to the homogeneous staple food group) but she can also purchase

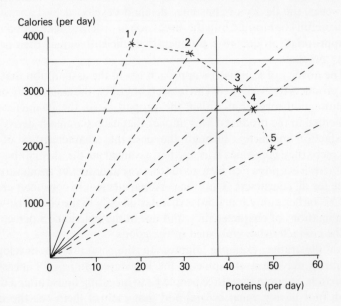

FIGURE 1.2   Illustration of Lancaster's theory with five meals including different amounts of calories and proteins

combinations of these in order to satisfy her family's needs for calories and proteins. This latter possibility could be indicated in Figure 1.2 by the line segments between points 1–2, 2–3, 3–4 and 4–5. Lancaster calls this set of segments the efficient substitution curve. Changes in the price of a product and in the income of the consumer can also be represented in the figure. For example, if the price of meal 1 increases, the new point 1 will move towards the origin along the line linking the latter and point 1. Similarly, if the income of one of the three people increases, he may choose to increase the amount he spends on a meal, so extending the efficient substitution curve.

Lancaster's theory has been discussed in some detail because it clearly illustrates several points that are important for the case-studies. First of all, it enables us to attempt to define appropriate products as those among alternatives, which given their price, satisfy basic needs most efficiently. Secondly, it is clear that in analysing the market demand for a given product it will generally be desirable to take into consideration its (close) substitutes. Thirdly, it shows that income (given a set of prices) is the main constraint on basic-needs satisfaction derived from consumption goods.

In the case of food products it is a relatively simple matter to identify the characteristics that satisfy basic needs. These characteristics usually include the amounts of calories, proteins, fats, vitamins, minerals, etc. contained in each product. In the case of non-food manufactured goods this identification is more difficult. In the case of shoes, for example, one might argue that the basic-needs characteristics include the consistency of the sole, the flexibility of the upper, the degree to which the shoes fit the foot and the shoes' durability. However, shoes may also satisfy needs other than those related to physical welfare, such as elegance and status (see note 10). Neo-classical economic theory makes no distinction between characteristics that satisfy basic needs and those that do not, taking the position that as long as there is an effective demand for a good, it increases the satisfaction of consumer needs (or utility). In contrast, the objective of satisfying basic needs implies an emphasis on the basic-needs characteristics of all products. The case-studies in this volume therefore attempt to identify these characteristics for the particular goods concerned, and to determine the extent to which the products produced by the various (labour- and capital-intensive) technologies have such characteristics.[12] Some of the case-studies also take into consideration the dynamic aspects of the efficiency with which goods satisfy basic needs. One of the reasons for increasing welfare in the developed countries, which is also valid (although to a lesser degree) in the developing countries, is that because of technological change the products now available to consumers satisfy a wider variety of needs, basic and non-basic, more efficiently. For the same price many products satisfy needs more efficiently; or, at a higher price, a given product satisfies a greater variety of needs. Taking this consideration into account, some of the case-studies presented in this book attempt to draw conclusions about the product design.

1.2 Appropriate Technology

Morawetz[13] defines 'appropriate technology' as ' . . . the set of techniques which makes optimum use of available resources in a given environment. For each process or project, it is the technology which maximises social welfare if factors and products are shadow priced.' Although Stewart[14] observes that the issue of the development of more appropriate techniques is implicitly ignored in this definition, it is a useful point of departure. The definition implies that a technically efficient technology is not necessarily the most appropriate one because it is the prices (and availabilities) of material inputs, skills, labour, capital and foreign exchange, together with the social welfare function, that finally determine the set of appropriate technologies. However, the definition does not specify the social welfare function which, for the purpose of this book, is defined as the satisfaction of basic needs within a generation (the approximate period of time suggested in the *Declaration* of the World Employment Conference). This objective can be attained in two ways: first, directly, by producing products that satisfy basic needs efficiently (i.e. at the lowest cost and incorporating the appropriate mix of basic-needs characteristics); and secondly, by providing employment and incomes for those whose basic needs are not fully satisfied. In the context of the case-studies in this volume this means that the whole range of technologies that produce products of different qualities have to be examined. The relevant characteristics of technology are, for example, the type of product, product nature, the scale of production, material inputs, labour and investment requirements.[15] After examining the characteristics of the technologies presently in use, some of the case-studies attempt to specify certain desirable characteristics of more appropriate technologies that could be developed in the future. Such characteristics include, for example, simplicity of operation and maintenance, the use of local materials, the production of appropriate products, etc.[16]

1.3 The Role of Marketing

The discussion of technological choice is often implicitly limited to production technology, excluding the marketing of the final product. This is not always justified, because there can be considerable variations in the labour-intensities of different marketing technologies. In addition, improved marketing systems could lead to greater production and thereby to more employment creation. This is particularly relevant for the informal sector, the products of which normally reach only the relatively small market of low-income consumers.

Other important aspects of marketing taken into consideration in the case-studies are the impact of advertising, price and the availability of appropriate products on the purchasing behaviour of consumers and on their basic-needs satisfaction. According to neo-classical economic theory, consumers ultimately determine the pattern of production and therefore individual purchasing behaviour is by definition rational. Section 1.2 on basic needs and appropriate products has already indicated that products which are appropriate to basic-needs satisfaction are unlikely to be abundantly available when there are serious income inequalities.[17] Such a situation will then lead to inefficient basic-needs satisfaction for the low-income groups. Another reason for system-wide inefficiency in satisfying the basic needs of low-income groups may be the influence of advertising which sometimes induces the consumer to make inefficient purchases. Inefficiency in the satisfaction of basic needs can be represented in a space with three dimensions: basic-needs and non-basic-needs characteristics and income. If the income pattern is projected on to the plane including the basic and non-basic-needs characteristics one may draw a graph as in Figure 1.3.

The segments $0A$ and AB which increase with income could be considered as the 'rational' buying behaviour, whereas the curve $0CD$ which also increases with income represents the actual behaviour, and the surface $0AC$ can be considered as the degree of 'inefficiency'. Examples of inefficient purchases are canned baby food where breast-feeding is better, cheaper and more hygienic; or hi-fi sets for households in which the basic needs for food, housing and education are not yet satisfied. The problem with the latter example is that the purchase of a relatively expensive hi-fi set may lead to benefits other than good music. It may also give the consumer a feeling of dignity and status which in some cultures may be of greater importance than the physical and physiological needs of the individual concerned. Although it is difficult to define precisely the concept of the degree of efficiency in buying

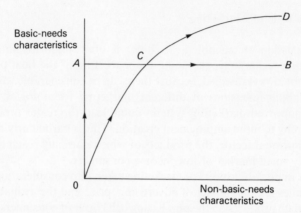

FIGURE 1.3   The 'hypothetical' and 'normal' path of basic-needs satisfaction
with increasing income

behaviour, policy-makers have to make a value-judgement about this in
order to formulate correctly their consumption policies. The case-
studies in this volume follow this policy approach and do not, therefore,
make a judgement about individual consumer choices.

A third aspect of marketing is the price of the product. In general,
price is a function of supply and demand. However, supply factors are
stronger than demand factors in some product markets. This is
especially the case when there is a limited choice of inappropriate
products, the production of which is in the hands of a few enterprises
having a strong influence on demand through advertising, a strong
dealer network or a production process offering significant economies of
scale. In such circumstances, price is more or less fixed and largely
independent of demand. Such situations may also be influenced by
government policies through either protective measures or excessive and
biased product-quality requirements. Taking such circumstances into
consideration, some of the case-studies in this volume attempt to analyse
the cost structure of products manufactured by the formal and informal
sectors and to identify policies that could encourage the production of
appropriate products at a price which can be paid by low-income
consumers.

1.4 Introduction to the Case-studies

The eight case-studies included in this book examine the relationships summarised in Figure 1.1 and further elaborated in the succeeding sections. The case-studies were completed between 1978 and 1981 and were largely based on surveys carried out by the authors themselves. Typically, each author carried out two surveys. First, they interviewed about 200–400 consumers (normally both in urban and rural areas) in order to collect data on income and purchasing behaviour. Secondly, they interviewed a number of enterprises (ranging from the very small to the largest) in order to analyse their cost structures and technology.

The order of presentation of the case-studies in this volume is as follows. In Chapter 2 we start with the study of Mubin and Forsyth on soap in Bangladesh and Ghana, since it gives a good overview of the various relationships in the analytical framework. Chapters 3 to 6 examine in detail the income distribution–technology–employment hypothesis and concern consumer durables. The authors, the product examined, and the country of analysis are respectively: Fong on bicycles in Malaysia (Chapter 3), Papola and Sinha on metal utensils in India (Chapter 4), Aryee on footwear in Ghana (Chapter 5), and House on furniture in Kenya (Chapter 6). This group is followed by the last three case-studies which mainly look into the basic-needs satisfaction that low-income consumers can derive from their given income. The authors, the product (or service) examined and the country of analysis are respectively: James on soap in Barbados (Chapter 7), Landgren-Gudina on weaning food in Ethiopia (Chapter 8), and Thobani on passenger transport in Karachi (Chapter 9). The final chapter (Chapter 10) attempts to summarise the conclusions of all the studies, also spelling out some of the more important policy implications.

Notes

1. All but one of the case-studies were published as mimeographed World Employment Programme research working papers. Their full titles are (in the same order as the chapters): A. Mubin and D. Forsyth, 'Appropriate Products Employment and Income Distribution in Bangladesh: A Case Study of the Soap Industry (Oct 1980): C. O. Fong, 'Consumer Income Distribution and Appropriate Technology: The Case of Bicycle Manufacturing in Malaysia (Mar 1980); T. S. Papola and R. C. Sinha, 'The Consumption Behaviour and Supply Conditions of Metal Utensils in India: A Study in the Basic Needs Framework' (1982); G. A. Aryee, 'Income Redistribution, Technology and Employment in the Footwear Industry (Jan 1981); W. J. House, 'Technological Choice, Employment Generation, Income Distribution and Consumer Demand: The Case of Furniture Making in Kenya' (May 1980); J. James, 'Product Choice and Poverty: A

Study of the Inefficiency of Low-income Consumption and the Distributional Impact of Product Changes' (May 1980): M.-A. Landgren-Gudina, 'Technological Choice, Employment Generation, Income Distribution and Consumer Demand: The Case of Weaning Food in Ethiopia' (1981); M. Thobani, 'Passenger Transport in Karachi: A Nested Logit Model' (1982).

2. Wouter van Ginneken and Christopher Baron are both staff members of the International Labour Office.

3. ILO, *Employment, Growth and Basic Needs: A One-World Problem* (Geneva, 1976) p. 31.

4. J. James and F. Stewart, 'New Products, a Discussion of the Welfare Effects of the Introduction of New Products in Developing Countries', *Oxford Economic Papers*, vol. 33, no. 1 (Mar 1981) pp. 81–107.

5. For example, W. Cline, *Potential Effects of Income Redistribution on Economic Growth: Latin American Cases* (New York: Praeger, 1972); D. Morawetz, 'Employment Implications of Industrialisation in Developing Countries: A Survey', *Economic Journal* (Sept 1974); V. E. Tokman, 'Income Distribution Technology and Employment in Developing Countries', *Journal of Development Economics* (Mar 1975); F. Paukert, J. Skolka and J. Maton, *Income Distribution, Structures of Economy and Employment* (London: Croom Helm, 1981).

6. Paukert, Skolka and Maton, *Income Distribution*, p. 104.

7. J. James, 'The Employment Effects of an Income Redistribution. A Test for Aggregation Bias in the Indian Sugar Processing Industry', *Journal of Development Economics* (Amsterdam: North-Holland, June 1980) pp. 175–89.

8. F. Stewart, *Technology and Underdevelopment* (London: Macmillan, 1977) in particular ch. III.

9. ILO, *Declaration of Principles and Programme of Action Adopted by the World Employment Conference* (WFC/CW/EI Geneva, June 1976).

10. Human beings are concerned (i.e. their 'utility' function is determined by) not only to increase their economic welfare (which is closely associated with basic needs) but also to enhance their physical security, their sense of participation in the society in which they live, and their feeling of identity within a family and/or cultural group. This last aspect of human behaviour causes low-income consumers in Ghana (see Chapter 5) to prefer elegance in footwear to durability, because of the importance of elegance in formal and ceremonial occasions.

11. K. J. Lancaster, 'A New Approach to Consumer Theory', *Journal of Political Economy* (Apr 1966) pp. 132–57. In his paper on sugar processing (Part I), James provided the link between Lancaster's theory and the concept of appropriate product. See J. James, *Technology, Products and Income Distribution: A Conceptualisation and Application to Sugar Processing in India* (Geneva: ILO, Nov 1977; mimeographed World Employment research working paper; restricted).

12. In this connection economic theory normally refers to products of different quality. In so far as the measurement of quality can be associated with clearly defined characteristics, no problem arises in applying the concept of quality instead of that of basic-needs characteristics.

13. Morawetz, 'Employment Implications of Industrialisation in Developing Countries: A Survey'.
14. Stewart, *Technology and Underdevelopment*, p. 95.
15. Ibid, p. 2.
16. See also ibid.
17. The surface BCD can be interpreted as the degree of over-satisfaction of basic needs, that is, surplus consumption of basic needs characteristics together with increased consumption of non-basic-needs characteristics.

2 Technology, Employment and Income Distribution: The Soap Industry in Bangladesh

A. K. A. Mubin and David J. C. Forsyth[1]

In this chapter we seek to examine how far the hypotheses set out in Chapter 1 are sustained by the evidence in the case of a particular product group, soap, in a typical low-income developing country, Bangladesh. Our principal interest is the analysis of the demand patterns for the various soap products in Bangladesh, their relationship to the choice of technology and employment, and the policy implications of these relationships.

The chapter is organised as follows. In section 2.1 income distribution, poverty and employment in Bangladesh are discussed to provide a general background to the study. Summary statistics on soap production and consumption are presented in section 2.2. Production technology is discussed in section 2.2.1 in theoretical terms and with regard to the specific nature of the soap manufacturing industry in Bangladesh. This is followed, in section 2.2.2, by an analysis of the nature and determinants of the demand for soap and a discussion of the significance of market imperfections. Section 2.3 considers calculations on the likely impact on employment in Bangladesh of income redistribution, and surveys the prospects for manipulating demand in such a way as to increase employment. Section 2.4 briefly summarises our findings for Bangladesh. Finally, an appendix compares the Bangladesh results with those of a similar study in Ghana.

2.1 Income Distribution, Poverty and Employment in Bangladesh

Bangladesh, one of the poorest countries in the world, has a population of approximately 85 million of which about 80 per cent are estimated to have incomes below the poverty line. There are no census data on the incomes of households or individuals in Bangladesh. However, data from recent sample surveys reveal that the lowest 20 per cent of households receive barely 10 per cent of income in rural areas and only 8 per cent in the urban areas, whereas at the other end of the income spectrum 20 per cent receive more than 45 per cent of income in urban areas and 40 per cent in rural areas. Available data suggest that income distribution in urban areas has tended to be rather less equal than in rural areas, though in recent years the respective figures have tended to move closer together.

Over all, income distribution in Bangladesh is less skewed than in other developing countries, a feature probably best explained in terms of Kuznets's inverted U-curve of distribution as it evolves over time. But an 'income-side' picture of shifts in distribution, which ignores price changes, may be misleading as a measure of economic well-being if such changes have different impacts on different classes of income earners. Khan has constructed a cost-of-living index for three groups – low, middle and high income – with 1963–4 as the base year. His figures[2] show that the index for the low-income group has risen much faster than that for the high- and middle-income groups; similarly for the index for the middle-income group *vis-à-vis* the high-income group. Thus the index rose by nearly 554 per cent from 1970 to 1975 for the poor as against a rise of 517 per cent for the middle group and 400 per cent for the high income group. Similarly, R. Islam found that prices of the items generally consumed by lower-income groups registered much faster rises than those of commodities consumed by richer groups. Thus these studies tend to indicate that the real standard of living of people in lower-income brackets has been affected more in relative terms by inflation than is evident simply from the distribution data presented above.

The magnitude of the problem of poverty in Bangladesh is indicated by the results of various studies of the provision of 'basic needs'. It is clear that the proportion of the population not provided with the minimum acceptable standards of diet, housing and clothing is staggeringly high and still increasing. As many as 60 million people, out of a population of 85 million, may be living below the 'poverty line'.

A significant indicator of poverty in the context of the present study is the extent of unemployment and underemployment. Comprehensive information on the employment situation in Bangladesh is not available, although it is known that rates of 20–47 per cent obtained in different towns in 1973. Dacca alone is believed to have some 300 000 beggars in its population of 1.6 million. The incidence of rural unemployment and underemployment is certainly high, but again hard to quantify. However, what is clear above all is that a vast reservoir of idle manpower exists – probably amounting to about one-third of the available labour force.

2.2 An Overview of Soap Production and Consumption

Soap is the oldest and most common detergent, and is known to have been used from earliest times in a variety of forms. In the more developed countries soap has largely been replaced by petroleum-based detergents for all but toilet use, but in the developing world it remains the most important cleaning agent.

Soap satisfies several basic needs and its use is widespread among the different income groups in Bangladesh. The soap produced locally may be classified into three categories: toilet soap, household or washing soap, and powdered soap. Toilet soaps are used for 'toilet uses' – that is, personal hygiene. These soaps come in various sizes and colours and are invariably attractively wrapped and scented. The weight of a piece of toilet soap varies between 3–5 ounces. Washing soap is purchased in ball or bar form (unwrapped) and is designed primarily for the washing of clothing although the 'average Bengalee' uses this soap for toilet purposes as well. Powdered soap is used only by laundries. It is sold on a limited scale to a specialised market, and is not therefore considered further here. For similar reasons, synthetic detergents are not discussed.[3]

2.2.1 Production

Soap is produced both by small producers on a cottage-industry scale and by modern large-scale factories. The former produce only washing soap. The latter produce both toilet and washing soap. In all there are fifteen modern mechanised factories in Bangladesh, with four in the public sector and eleven in the private sector. The largest single enterprise in the private sector is a subsidiary of Unilever. All the large-scale producers are concentrated in the two most important cities, Dacca and Chittagong. Small-scale factories, on the other hand, are scattered all over the country. According to the *Directory of Industries*, 273 such units were operating in various towns in August 1978, all privately owned. Accurate statistics on employment in the industry are not available, but a figure of about 5000 seems plausible.

It is estimated that the total production capacity installed is about 60 000 tons annually. However, because of a shortage of raw materials the industry is operating well below capacity and, according to one estimate, only 20 000–25 000 tons are being produced annually. Toilet soap accounts for about 20 per cent of all soap production and the balance is mainly washing soap.

Government policy towards the soap industry. Prior to the emergence of Bangladesh as a separate state, the larger manufacturing units in the soap industry, in common with other manufacturing industry, expanded under a regime of heavy tariff protection, an over-valued exchange rate, and ready access to finance – with the result that many enterprises developed relaxed attitudes. In the rush for industrialisation in the modern sector traditional and small-scale industries were starved of essential supplies of materials, and were given low priority in the use of electricity, fuels and transport facilities. They had little access to the import licensing authorities and had to rely on intermediaries, paying higher prices for such materials as they could acquire.

After the liberation of Bangladesh in 1971 the ensuing confusion led to a decline in annual industrial production of the order of 30 per cent. The shortage of 'domestic' raw materials, owing to the disruption of supplies from Pakistan together with a sharp rise in the price of imported raw materials, affected the soap industry considerably. In order to offset the scarcities and consequent increases in product prices, new policies were adopted to encourage the maximum utilisation of existing capacity. The soap industry was granted import licences for raw materials in excess of 100 per cent of its guideline quota and efforts were made to improve the delivery of vital raw materials by improved planning. Furthermore, heavy emphasis was placed on the improvement of management methods and labour relations in the public sector.

These, then, are the principal features of government policy in so far as it has impinged on the soap industry in recent years. It is important to note that no policy measure has been aimed specifically at the industry although it has been affected by policies directed at manufacturing as a whole.

Alternative production techniques. Labour-intensive factories use a simple soap-making process. Saponification[4] is done by heating the mixture of fats and oils with caustic soda in a pan. The heating time varies from twelve to seventy-two hours, depending on the quality of the soap to be produced. After heating, the mixture is allowed to cool and the thickened soap poured into frames or pots for sizing and drying. Most soap factories add sodium silicate to the soap mixture, and prepare the silicate by heating in a separate pan. If desired, colour or scent may be added at the cooling stage. This type of process requires only pans for boiling, and frames of wood, iron or steel for moulding the soap as it solidifies. In some cases a cutting machine is used for simultaneously cutting and stamping the soap. In the typical factory the following stages of production can be identified: (i) raw material collection and storage, (ii) melting of sodium silicate in a pan by boiling, (iii) saponification, that is, mixing and heating the various materials (oils, fat, liquid caustic soda, prepared silicate) from twelve to seventy-two hours, (iv) cooling of thickened mixture – and pouring into frames, (v) storage until solid and dry, (vi) cutting (optional), (vii) stamping (optional) with the help of a stamping machine or manually with a seal, and (viii) packing in boxes (usually wooden).

Labour-intensive factories are usually small-scale operations producing washing soap only. The capacity of a process pan may vary from 6–20 maunds (1 maund (md) = 82.29 lb). Materials handling is always manual. The fuel used for heating is usually firewood, but where gas is available it is preferred.

Capital-intensive factories apply essentially the same techniques for making soap, but have facilities for bleaching and washing the raw materials, for separating soap and glycerine, for skimming and pumping neat soap from brine and water, and for drying and kneading the soap in order to achieve uniformity and homogeneity of texture. Machinery is also used for drying, spraying, cutting and stamping. Materials are pumped through pipes from one stage to another.

The various stages in the production process are: (i) receipt and storage of raw materials (which come in liquid, semi-liquid or solid forms) in drums, bowsers, bags, tanks etc., (ii) melting of solid and semi-liquid materials and bleaching where necessary, (iii) saponification,[5] (iv) glycerine extraction, (v) fitting and settling for about thirty-six hours (a homogeneous mixture of caustic, brine and water), followed by skimming and pumping of soap to a waiting vessel, (vi) drying – by raising temperature and spraying, (vii) chemical mixing, (viii) 'plodding', or kneading the soap into solid bars, (ix) cutting into suitable sizes and 'weathering' (i.e. secondary drying), (x) stamping, (xi) wrapping (toilet soap only), and (xii) packing.

Stages (iii)–(ix) may be regarded as the production process proper, (i) and (ii) the materials receipt and preparation phases, and (x)–(xii) the finishing and warehousing phases. In the materials-preparation stage a typical large-scale capital-intensive factory has separate units for melting tallow and coconut oil, dissolving the caustic soda, and bleaching to furnish high-quality final products. (For other products inputs are not bleached.) After preliminary preparation the raw materials are stored in storage tanks to be used whenever necessary. Where bleaching is done these factories discard 'spent earth' which contains about 16 per cent oil and is sold to produce low-grade soap.

The production stages largely involve automatic operations. The materials are pumped from stage to stage and only a few operators are needed to control the flow. These operators need skill and experience and work under the guidance of a supervisor who possesses expertise and some formal education.

In the finishing section a large amount of labour is required. Activities like carrying finished products from one place to another, packing into boxes, box making, carriage and warehousing require large numbers of unskilled men, although trollies and fork-lift trucks are sometimes used.

A modern soap factory must be housed in a much bigger building than that required by the labour-intensive factory. In general, the production processes are located on an upper floor and the finishing activities on a lower one, so the structure must be purpose-built.

Detailed characteristics of the alternative technologies used in Bangladesh. The fieldwork in this study involved a detailed investigation of the techniques for making soap in Bangladesh. A questionnaire was completed in the course of visits to factories on the basis of discussions with managers, engineers, accountants and proprietors. Basic information on labour, capital, raw material requirements and product quality was collected, as were particulars of capacity utilisation, marketing, problems of material input and skill availability, etc.[6]

The survey covered sixteen units: fourteen in the labour-intensive sector and two in the modern large-scale sector. (In fact, four large, capital-intensive factories were identified and interviewed. However, one had suspended production some time earlier and another provided only limited information.) As regards the labour-intensive sector, a complete list of producers was obtained from the Directorate of the Ministry of Industries. Limited time and resources constrained the survey of these firms, but the final tally seemed satisfactory given the fairly homogeneous nature of this group. The results of the survey are presented and discussed below.[7]

An examination of the labour input in the sixteen factories revealed that in all but one of the labour-intensive establishments a single-shift was worked whereas the capital-intensive factories operated a multiple-shift system. However, the range of values observed was again very considerable. The smallest number employed was eight persons and the largest 981. All the factories employed more direct than indirect labour.

Using the above data on output, investment and employment we can generate the figures presented in Table 2.1. Once again we find varying values within the labour-intensive sector. Capital-intensity, as measured by expenditure on plant and machinery per unit labour, ranges from Tk1091 to Tk11 364 per man, and there appears to be a weak positive association between scale and capital-intensity in this group. However, the two large factories (nos. 15 and 16) are by far the most capital-intensive in the sample and clearly constitute a separate sector: firm 16 employs over one hundred times more capital per man than does firm 1. A different picture emerges when we use an alternative measure of capital-intensity, the capital–output ratio. Here there is, if anything, a tendency for the ratio to decline as output rises in the labour-intensive

TABLE 2.1　Main characteristics of the factories in the survey

Factory number†	Capital*/labour		Capital*/output (Tk/md)	Labour†/output (employees/000) (md)
	Direct labour	All labour		
	(Tk/man)			
1	1 500	1 091	66.7	61.1
2	5 000	3 333	125.0	37.5
3	3 333	2 500	50.0	20.0
4	4 286	3 000	36.6	12.2
5	6 250	5 000	58.8	11.8
6	6 500	4 727	52.0	11.0
7	5 000	4 167	46.3	11.1
8	4 167	2 778	33.3	12.0
9	7 917	5 588	47.5	8.5
10	15 625	11 364	52.1	4.6
11	3 500	2 333	14.0	6.0
12	4 667	3 333	20.0	6.0
13	5 555	3 704	13.3	3.6
14	4 000	3 509	10.6	3.0
15	44 872	35 678	152.8	4.3
16	209 412	138 927	208.7	1.5

* Plant and machinery, replacement cost.
† All labour.

sector, although the values for the two large firms are again well above the rest. Finally, the labour-to-output ratio, which indicates the relative productivity of labour (lower values indicating higher productivity), behaves as might be expected, with labour productivity tending to rise with the capital–labour ratio.

What are we to make of these results? Within the smaller-scale, labour-intensive sector (factories 1–14) low outlays on plant and machinery per man appear to be associated with low productivity of both capital *and* labour. Thus very small, labour-intensive plants may be 'technically inefficient', and the interaction of increased scale of operations and increased mechanisation probably yields significant economies of scale. Particularly high capital–output ratios are recorded for the two large plants, but the labour productivity measures tend to suggest that here we are on familiar neo-classical territory, with a trade-off between factor productivities as factor proportions are varied.

These rather surprising results may be viewed from a different angle if we ask what would happen to input requirements if all soap production of our sample were to be carried out using only one of the techniques listed in the table? Assuming that all the techniques display constant returns to scale, we can answer this question by multiplying the capital and labour inputs for each technology (i.e. each factory) by the inverse of the proportion of total output accounted for by it. The figures so generated are presented in Table 2.2, which clearly reveals the extent of apparent 'technical inefficiency' in the sample of sixteen techniques. While no wholly 'dominant' technique emerges, only three variants are 'efficient' (nos. 14, 15 and 16).

Such an approach to the comparison of techniques is of course very crude. Apart from assuming away economies of scale, it assumes zero (or equi-proportional) excess capacity, no substitutability between capital, labour and other inputs, identical products, and 'correctly' priced capital goods. In fact, all these assumptions are unlikely to be valid, so that we cannot legitimately conclude that only these three techniques of production should be considered. However, it seems fair to reiterate the tentative conclusion that there may be significant diseconomies of scale in small-scale production which is likely to be less attractive in commercial terms than the larger units. At the same time, the capacity of the former for creating employment is evidently considerable. Stated summarily, the entire output of the sixteen plants covered could be produced by two factories similar to no. 16, creating 820 jobs, or by 3069 small units employing 33 759 workers. The capital requirement for the large-scale approach (considering plant and ma-

TABLE 2.2 Input requirement for alternative methods of producing total output of sample firms*

Factory type	Number of units required	Capital (Tk '000)		Labour (number)
		Plant and machinery	Total	
1	3 069	36.8	98.2	33 759
2	2 301	69.0	103.5	20 709
3	1 381	27.6	41.4	11 048
4	674	20.2	NA	6 740
5	650	32.5	84.5	6 500
6	552	28.7	45.3	6 072
7	511	25.6	NA	6 132
8	368	18.4	23.9	6 624
9	276	26.2	45.5	4 692
10	230	28.8	40.3	2 530
11	221	7.7	40.9	3 315
12	158	11.1	26.9	3 318
13	37	7.4	9.3	1 998
14	29	5.8	11.6	1 653
15	2.4	84.0	114.0	2 354
16	2.0	113.9	154.9	820

* Aggregate output = 552 330 maunds.
NA = not available.

chinery alone) would be Tk114 m and for the small factories Tk37 m. Clearly, the industry could not in fact be organised entirely on a labour-intensive basis in reality unless there were to be radical changes in the product-mix and consumer tastes, but the figures usefully illustrate the orders of magnitude involved.

From the point of view of investment strategy, economic efficiency (rather than technical efficiency) is the more immediately relevant consideration. Are labour-intensive plants more, or less, costly to operate than capital-intensive ones? This was examined in the case of soap by comparing costs and revenues in four of the labour-intensive plants known to have been operating at or near full capacity with the two capital-intensive plants.

For each of these factories projections of all cash flows were made over a twenty-five year period, and the net cash flows then discounted at discount rates ranging from 10–30 per cent per annum. All plant, machinery and buildings were valued at replacement cost. Labour, raw materials, and other current inputs were valued at base year prices (no allowance being made for inflation). Capacity utilisation was assumed to be 30 per cent in the first year of operation and 100 per cent of declared sustainable capacity thereafter.

The net present value (NPV) figures for the six factories are shown in Table 2.3. Results are given for the actual factory sizes, and in the cases of labour-intensive cases *A*, *B*, *C* and *D*, pro-rated for output equivalent to the larger of the two capital-intensive factories, case *E*. It will be seen that the capital-intensive case *E* is clearly more profitable (i.e. has a greater positive NPV) than any of the labour-intensive variants at all discount rates except in the case of factories *A* and *B*, when the 30 per cent discount rate is used. However, this result is more encouraging for protagonists of the labour-intensive technique than might first appear to be the case because (i) scale economies are ignored in cases *A*, *B*, *C* and *D*, (ii) 'shadow' discount rates are high in most developing countries and a rate of 30 per cent may be more appropriate than one of 10 per cent, (iii) 'shadow' wage rates will be well below observed wage rates, and this will favour *A*, *B*, *C* and *D*, (iv) employment generation is considerably greater in *A*, *B*, *C* and *D* than in *E* and *F*, and some social return attaches to this, and (v) the results for plant *F* are not clearly superior to these for *A*, *B* and *C* although *F* is a successful, competitive firm.

Thus the figures given in Table 2.3 suggest that labour-intensive technologies, especially of the kind embodied in factories *A* and *B*, may be just as attractive as capital-intensive technologies for the production of washing soap. This examination has not been exhaustive, and further

TABLE 2.3 NPV figures for six soap-producing factories

| Plant | Capacity (tons per annum)* | NPV (Tk million) | | | | | |
| | | At actual capacity | | | Pro-rated to equal output of plant E | | |
		10%	20%	30%	10%	20%	30%
Labour-intensive							
A	550	5.48	2.49	1.33	91.33	41.50	22.17
B	690	6.89	3.08	1.59	91.62	40.96	21.14
C	128	0.94	0.23	(0.04)	67.14	16.43	(2.86)
D	92	(0.12)	–	–	(12.00)	–	–
Capital-intensive							
E	9 174	231.5	72.1	21.9	–	–	–
F	8 073	116.8	27.3	(7.7)			

* Correct to nearest ton.
( ) indicates negative NPV (loss).

detailed analysis of the economics of the technological alternatives is needed before advice can be offered on investment decisions with any great confidence; but at least it is clear that labour-intensive methods of production cannot be ruled out as uncompetitive.

In addition to these considerations, as indicated earlier, the equipment required by the alternative techniques differs considerably, with the capital-intensive method involving more sophisticated processes, and consequently requiring more complex machinery. Table 2.4 lists the actual equipment used in Bangladesh and the source of supply in each case. It shows that the labour-intensive approach uses virtually no directly imported items (although certain components and some of the iron and steel may have been imported) whereas the capital-intensive system depends heavily on imports.

Raw materials requirements and the sources of supply. The two techniques also differ in their use of raw materials and sources of supply of these materials. Table 2.5 shows the sources and uses of inputs of the two techniques. The table shows that important raw materials like tallow, coconut oil and caustic soda are mainly imported. Manufacturers complain that local tallow and oils are of poor quality and difficult to obtain. However, it is believed that most of the raw materials (particularly animal and vegetable oils and fats) could be produced on a regular, reliable basis in Bangladesh – with obvious benefit to both using and supplying industries.

It is clear from Tables 2.4 and 2.5 that the labour-intensive technique requires many local inputs, of raw materials and from engineering firms. On the other hand, the capital-intensive technology depends heavily on imported raw materials, plant and equipment. Thus, if we consider the spread effects of the choice of technology on the Bangladesh economy, these are likely to be much greater in the case of the labour-intensive technology.

The physical and chemical properties of the soap produced depend on the nature of the raw materials used in the production process, and on the nature of the process itself.

Physical properties such as the dissolving characteristics of soap are greatly influenced by the fats (such as tallow) used. Generally speaking, a high fat content produces hard soaps which dissolve slowly but give a stable lather. A high oil content produces soft soaps that dissolve quickly. The degree of susceptibility to oxidisation and hence to rancidity again depends on the fats used: for example, coconut oil is resistant to oxidation. The quality of fats used also affects colour, as

TABLE 2.4 Equipment required and source of supply for the alternative techniques

Production stage	Capital-intensive technology Equipment required	Source of supply	Labour required technology Equipment required	Source of supply
Unloading of raw materials	Drums, trucks, bowsers, bags, trollies, fork lifts, pumps, pipes	Imported/local	Push cart, head-load bags, trucks, drums	Imported/local
Storage	Tanks, bags	Imported/local	Bags, tanks	
Treatment of raw materials	Weighing scale	Imported	Weighing scale	Imported/local
	Melting trench,	Imported/local		
	Bleacher dissolving unit	Imported/local	Pan	Local
	Steam heating pan	Imported/local		
	Pumps, pipes, steam nozzle	Imported/local		
	Press fitted with synthetic cloth	Imported		
Saponification	Pan and kettle	Imported/local	Pan	Local
Filling and settling	Stock pan, pumps	Imported/local	Pan	Local
Glycerine extraction	7-stage divided pan unit	Imported	–	–
Chemical mixing	Mixer, crutcher	Imported	–	–
Drying	Vacuum chamber	Imported	Iron or wooden frames, small pots, etc.	Local
	Heat exchanger	Imported		
	Spray dryer	Imported		
Kneading	Plodder	Imported	–	–
Cutting and weathering	Cutter	Imported	Iron thread	Local
			Simple cutter	Local
Stamping	Stamping machine	Imported	Seal	Local
			Simple stamping machine	Local
Wrapping	Wrapping machine	Imported		
Packing	Wooden boxes	Local	Wooden boxes	Local
	Cartons	Local		
	Trollies	Local		

TABLE 2.5 Input requirements of the alternative techniques of soap production and sources of supply

Technique	Raw materials	Source of supply
1 Labour-intensive	Tallow	Local/imported
	Coconut oil	Imported
	Indigenous oils (nut oil, mustard oil, etc)	Local
	Sodium silicate	Local
	Caustic soda	Local/imported
	Soap stone powder	Imported
	Colour	Local
	Packing materials	Local
	Firewood	Local
	Glass	Local
2 Capital-intensive	Tallow	Imported
	Coconut oil	Imported
	Caustic soda	Imported/local
	Perfumes	Imported
	Colour	Imported
	Bleaching agent	Imported
	Packing materials	Local
	Furnace oil	Imported
	Electricity	Local

poor-quality fats contain extraneous materials which discolour the final product whereas pure, refined fats and oils form white soap. Again, the type of fats used affects the 'mildness' of the soap; tallow soaps are softer to the skin than those made from other fats. The quality of the bases (i.e. alkalis) used similarly affects the physical properties of soap. Generally speaking, soaps made from sodium are firm and slow to dissolve, whereas soaps incorporating potash and ammonia tend to be soft and pasty.

Capital-intensive factories in Bangladesh normally use only imported tallow and coconut oils. These raw materials are of superior grade and produce good-quality soap. On the other hand, the small-scale labour-intensive factories make use of tallow and oils, in addition to imported materials, and so the product is generally of lower quality. This seems to be inevitable given the present nature of the labour-intensive sector, as it is difficult for small, dispersed producers to gain access to supplies of imported material on a regular basis. However it is not an inherent weakness of labour-intensive technologies, because they could use high-quality materials if these were available. What is true is that where impurity-free raw materials are not available then impurities in the raw materials must be removed before good-quality (i.e. firm, mild, stable) soaps can be produced. Here the capital-intensive factories have a clear technical advantage over their labour-intensive rivals, in virtue of their facilities for bleaching and cleansing. Labour-intensive units normally cannot do this, and the quality of their product suffers accordingly. Their soaps are markedly less mild and produce a poorer lather than 'capital-intensive' soaps.

Moreover, in the small labour-intensive factories glycerine cannot be extracted; this affects the character of the soap, and means that a valuable by-product is lost. In the capital-intensive factories glycerine extraction and separation equipment is installed.

In the production of toilet soap, a lower moisture content must be achieved than is the case with washing soap. This is achieved through a special 'milling' process – a series of drying, spraying and vacuum processes designed to reduce the water content. In labour-intensive production, toilet soap can be made by the cold process, but this has two important disadvantages: (i) some unsaturated fats or free alkali often remain in the soap, and this is undesirable in toilet soap, (ii) there is no simple way of cooling and drying so that perfumes and colour have to be mixed when the soap is still warm, causing the more delicate (and expensive) perfumes to evaporate. Thus the simple process is not suitable for the production of good-quality toilet soap. For these reasons the production of toilet soap is confined to the capital-intensive sector in Bangladesh.

The outcome of these relationships between materials, technology and product quality is that three distinct soaps are manufactured in Bangladesh. Toilet soap is only produced by the capital-intensive factories, but washing soap is produced by both capital- and labour-intensive techniques. In the light of the issues already mentioned, it is to be expected that the former method of production will produce superior washing soap, but the comparative disadvantage of the latter is not such as to eliminate it from the market.

2.2.2 The consumption of soap

If household demand for soap varies systematically with income then changes in the distribution of income will affect the total demand for soap. If the kind of soap demanded also depends on income, then changes in income distribution may affect the relative demands for different types of soap. With certain caveats, it is possible to forecast the likely impact of changes in income distribution over time by analysing cross-sectional data on consumption. Data were obtained in this study by a detailed survey of consumption habits in Bangladesh. The results are presented and discussed below. We then examine the purpose for which soap is purchased by the different income groups, the characteristics of the product most relevant to these purposes, and the extent to which the characteristics of the types of soap purchased match the requirements of the user.

The consumption survey. A survey of purchasing behaviour was carried out in samples of rural and urban households by face-to-face interviews based on a questionnaire.[8] The questionnaire included questions on the various uses of soap: the type and quantity purchased per month, the frequency of purchase, and so on. Respondents were asked to identify and rank the attributes or characteristics that they regarded as desirable in soap, given the uses to which it was put by them.

The sample structure was designed to facilitate analysis of the impact on demand patterns of two factors that were expected *a priori* to be of particular importance – income levels and place of residence (in terms of a rural–urban dichotomy). Table 2.6 shows that in both rural and urban areas average household expenditure on soap increases with income indicating that, over the range of income levels in Bangladesh, soap is not an 'inferior' good.

TABLE 2.6   Consumption of soap by income group, rural and urban areas, and for country as a whole

Household income (Tk/month)	Households in sample			Average monthly purchase of soap (Tk)			% of income spent on soap		
	Rural	Urban	Total	Rural	Urban	Total	Rural	Urban	Total
0–99	1	–	1	–	–	–	–	–	–
100–149	1	–	1	1.50	–	1.50	1.25	–	1.25
150–199	5	–	5	5.40	–	5.40	3.50	–	3.50
200–249	14	–	14	7.76	–	7.76	3.80	–	3.80
250–299	6	–	6	8.66	–	8.66	3.24	–	3.24
300–399	30	1	31	10.59	7.50	10.49	3.70	2.14	3.64
400–499	17	8	25	13.96	14.49	14.12	3.60	3.62	3.60
500–749	20	27	47	17.39	19.66	18.69	3.20	3.19	3.19
750–999	6	30	36	28.00	27.88	27.90	3.40	3.87	3.79
1000–1499	4	39	43	49.24	35.92	37.15	3.80	3.28	3.32
1500–1999	4	12	16	56.49	56.21	56.30	3.50	3.33	3.37
2000 and over	2	33	35	47.75	77.76	76.04	2.30	2.06	2.07
	110	150	260						

Both rural and urban areas yielded a maximum figure for soap consumption as a proportion of income of some 3.8 per cent, with a falling away to well below 3 per cent for monthly incomes of Tk2000 and above. Perhaps the most striking characteristic of the consumption/ income relationship is the fairly narrow range in which most of the values fall.

The aggregate figures for soap consumption out of income conceal widely divergent trends at the sub-product level. All the soap consumed in Bangladesh falls into one of three clearly defined categories – toilet soap, washing soap made by a capital-intensive process, and washing soap made by a labour-intensive process – and the analysis may be refined if we examine the consumption of each separately.

In Table 2.7 average monthly expenditure on each of these types of soap is shown for each income class. A clear pattern of market segmentation emerges from the figures. Labour-intensive washing soap is purchased mainly by the lowest income earners. As we move up through the income brackets capital-intensive washing soap, and then toilet soap, becomes important. The clear implication of these data is that labour-intensive washing soap tends to be bought by the poorer sections of the community. Although there is some 'brand allegiance',

TABLE 2.7   Distribution of soap consumption by location/soap type

Monthly household income (Tk)	Average monthly purchase of washing soap per household				Average monthly purchase of toilet soap per household (Tk)	
	Labour-intensive		Capital-intensive			
	Rural	Urban	Rural	Urban	Rural	Urban
0–99	0	–	0	–	0	–
100–149	1.50	–	0	–	0	–
150–199	5.40	–	0	–	0	–
200–249	5.57	–	0.43	–	1.25	–
250–299	6.00	–	2.67	–	0	–
300–399	8.33	0	2.03	7.50	0.22	0
400–499	8.11	1.25	3.53	6.69	1.64	2.76
500–749	8.40	4.46	7.60	9.21	1.40	5.26
750–999	6.00	2.32	15.67	19.99	2.00	10.23
1000–1499	26.50	1.00	3.50	23.36	20.25	14.21
1500–1999	39.00	2.00	10.50	30.25	7.00	23.69
2000 and over	6.00	0	24.00	47.48	17.75	28.00

– indicates not covered in sample.

this cannot overcome the tendency towards the purchase of capital-intensive washing soap, and then toilet soap as incomes rise. Particularly striking is the fact that of thirty-five cases in our sample with income of Tk2000 (and over) per month, only one consumed the lower-grade washing soap, all the others consumed the higher-grade products.

Price and the demand for soap. The prices of the different types of soap sold in Bangladesh depend on the technique of production, but much less so within individual product groups. In particular, the average retail price of washing soap made by a capital-intensive process was found to be Tk7.6/lb as compared with Tk3.5/lb for washing soap made by a labour-intensive process. The prices of toilet soaps varied according to quality and type, ranging from Tk14/lb to Tk24/lb. For each type of soap unit prices varied, but not sufficiently to invalidate the usual assumption of price uniformity within product groups adopted in the econometric analysis to be presented below.

However, it is possible to make some impressionistic comments on the influence of relative price variation on the demand for each product. First, it seems unlikely that the demand of consumers in the upper-income brackets for labour-intensive washing soap is price-elastic. Given the strength and uniformity of the swing to the more expensive capital-intensive variant as incomes rise, labour-intensive producers may anticipate some difficulty in inducing a swing back to their product by those already purchasing the capital-intensive soap simply by reducing price, although such price reductions might have a favourable effect on demand for the product from lower-income consumers. However, the precise magnitude of this last effect cannot be satisfactorily estimated from the survey data. At the same time, it is conceivable that labour-intensive washing soap may have the 'Giffen good' property, that the negative real-income effect of a price *rise* might compel some consumers to reduce their purchases of the more expensive soaps and substitute labour-intensive washing soaps.

Again, any increase in the prices of the capital-intensive products is, other things equal, likely to raise the income threshold for switching to these products from the labour-intensive variants, and motivate some consumers to revert to labour-intensive soaps. Equally, any reduction in the prices of capital-intensive soaps may accelerate the switch away from the labour-intensive product.

Modelling the consumption relationship.[9] In order to examine the determinants of aggregate soap consumption more systematically, a formal model of the demand relationship was constructed. It was expected that the variation across households in per capita consumption of soap could be explained partly in terms of differences in income, but other significant factors could be the size and composition of the household and the social context, location, etc. It was also anticipated that the pattern of demand for different kinds of soap would diverge, and an attempt was made to examine the determinants of demand for each of the main kinds of soap separately.

In this section per capita (rather than per household) consumption is used as the unit of measurement in conformity with standard practice in household budget analysis. This facilitates the analysis of the influence of variations in household size on the level of consumption, making possible the separating out of 'economies of scale in consumption'. Although average household size increases with household income in both rural and urban areas, this effect was found not to be strong enough to affect perceptibly the rankings of cases by individuals or by household groupings respectively. In other words, no clash is expected here, and the results for by household and per capita analysis are believed to be interchangeable.

The *a priori* considerations noted above may be expressed precisely by:

$$C = f(I, H, J, P) \qquad \text{equation (1)}$$

where C = per capita consumption of soap (in each household); I = per capita income; H = number of members; J = occupation of head of household; and P = location of residence.

It was anticipated that the relationship between consumption and income (both measured in Tk/capita) would follow the characteristic Engel curve pattern, and several alternative versions of this equation were examined for 'explanatory' power and goodness of fit. Although it seemed reasonable to adopt the 'homogeneity hypothesis' with regard to the composition of the family (i.e. all members are assumed to consume the same amount of the product irrespective of age) the possibility of scale economies remains. An attempt was made to capture this possible effect by the inclusion of the household size variable, H. It was also anticipated that soap consumption would tend to be higher for white-collar occupation ($J = 1$) than for other workers ($J = 0$) and for urban dwellers ($P = 1$) than for rural dwellers ($P = 0$).

All the functions estimated took the general form of equation (1), and ordinary least squares was used throughout. Experiments with alternative functional forms suggested that a simple linear relationship performed best for both aggregate consumption of soap and consumption of different soap types.

The demand for soap.   The results of the analysis are presented in detail elsewhere, but the main findings may be summarised as follows: per capita consumption of all soaps was found to be higher in urban than in rural areas; the factors associated with greater levels of consumption were the same in both cases – higher income (which constituted the strongest influence on consumption), smaller household size, and non-manual occupation. This finding was repeated in the case of individual soap types, though there were important differences *between* soaps. Taking all consumers together, urban dwelling was found to be associated with greater consumption of the more sophisticated soap, but with diminished purchases of labour-intensive washing soap. Consumption was positively related to income (the dominant regressor) for both of the capital-intensive soaps and for the labour-intensive washing soap (although in the latter case the relationship was not significant at the 5 per cent level). For the two capital-intensive soaps, household size exerted the anticipated negative effect on per capita expenditure (indicating the presence of economies of scale in consumption) though no such effects were detected in the case of labour-intensive washing soap.

Finally, the impact of occupation was highly significant, although in opposite directions, in the cases of toilet soap (where manual work was associated with higher per capita consumption), and capital-intensive washing soap, the demand for which was greater among non-manual workers.

When the sample was broken down into its two geographical components, the main changes from the above configuration of results were that: (i) in the case of toilet soap the occupation variable was non-significant, (ii) in the case of capital-intensive washing soap (rural consumers only) the household size variable was non-significant, (iii) in the case of labour-intensive washing soap, splitting the sample yielded a non-significant relationship (in terms of an over-all F-test) for the 'rural' equation as a whole, but did bring out the relationship anticipated *a priori* between possession of a manual job and increased consumption of this kind of soap.

The foregoing analysis was carried out on the assumption that, for all soaps and for all consumers, 'other things were equal'. However, there is one particular exception to this rule which it would be unwise to ignore and about which we do have specific expectations: because of the pressures of advertising, the 'demonstration effect' and so on, purchasers may often buy a more sophisticated variant of a product than they actually need. In the present context this would mean the purchase of an excessively sophisticated soap – one that possesses 'superfluous' attributes and characteristics. If, on analysis, the qualities *sought* in soap do not correspond to the properties *incorporated* in the soap purchased, then more efficient purchasing behaviour may be worth considering. Of particular interest in the present context is the possibility that such 'errors' on the part of purchasers of soap may lead them to buy capital-intensive soap where the labour-intensive variant would suffice, with consequent adverse effects on employment generation and the balance of trade.

In this section we look first at the uses to which soap is put. We then present the results of our investigation of the properties desired by soap consumers relative to particular uses of soap. The actual properties of three main kinds of soap are then discussed, and an attempt made to assess whether or not the matching of soap purchases to soap use is clearly rational, or perhaps suggests the presence of some 'property illusion' on the part of the buyer.

Our survey revealed that soap is used for two main purposes – the washing of clothes ('laundry' use) and personal hygiene ('toilet' use). Other uses were found to be unimportant and the subsequent discussion deals only with these principal uses. All 260 households visited reported that they used soap for laundry purposes, and only twenty-eight reported no use of soap for toilet purposes. Thus soap is used for two purposes, of which the 'laundry' use is clearly the primary use in all but the highest-income households. For all income levels up to Tk2000 per month almost all respondents rank laundry use first. Although, as income increases, there is a growing tendency to make use of soap for toilet purposes, this is regarded as a secondary use. Significantly, the position in the highest income group is dramatically different, with the 'toilet' use of soap being the primary one in a majority of households.

We now examine the demand for the inherent properties of soap, that is, the properties that are perceived to be important by users given the purpose to which they put soap; what properties are desired by consumers who use soap primarily for laundry or toilet purposes?

Soap has various attributes that correspond to the particular needs or preferences of the purchasers. In the course of our consumption survey we identified the following as important: washing ability, durability, resistance to breakage, scent, colour and wrapping.

Washing ability depends on the chemical properties of soap as a cleaning agent, and encompasses factors such as its ability to form a good and stable lather and to minimise damage to clothing, and its 'mildness' to the skin. Durability and resistance to breakage are straightforward; some soft soaps have the disadvantage of washing away quickly, while others do not; again, some soaps exhibit 'bunching', that is, formation of separate grains in the mix which makes such soap less resistant to breakage and consequent wastage. Otherwise scent, wrapping and colour are important to various users.

Which of these attributes are likely to be of interest to the respective user groups? What particular 'clustering' of desirable attributes is reflected in the demand for properties on the part of the respective groups? In order to probe this important point, respondents were asked to rank the various properties listed above in terms of their perceived desirability given the use to which the soap was to be put. In Table 2.8 the ranking by respondents of the attributes desired in soap for laundry and toilet use is indicated. First, second and third rankings were requested in order to focus attention on the key attributes.

Looking first at the data on soap used for laundry purposes, we find that 90 per cent of all respondents ranked 'washing ability' as the most important attribute of a soap to be used in this way. A further 8 per cent regarded 'durability' as the main attribute, and 1 per cent nominated resistance to breakage. Among the attributes ranked second, durability and resistance to breakage were the most frequently mentioned, though wrapping and scent were now regarded as important by nearly 9 per cent of all households (or 13 per cent of respondents). Among the responses from the few households offering a third ranked property, scent was the most frequently mentioned. Colour was not mentioned. It seems reasonable to sum up these responses as being heavily biased towards severely 'practical' rather than aesthetic aspects.

TABLE 2.8   Ranking of characteristics desired in soap
(% of households)*

Ranking	Washing ability	Durability	Scent	Wrapping	Resistance to breakage	Colour
			Soap for laundry use			
1	90.20	8.25	–	–	1.18	–
2	4.31	48.23	2.35	6.27	7.05	–
3	–	0.39	4.31	2.35	3.92	–
			Soap for toilet use			
1	52.46	13.50	32.70	0.61	0.61	–
2	28.40	29.00	24.00	3.08	8.60	0.61
3	7.40	11.00	11.00	12.30	3.00	1.20

* Not all households gave second and third rankings; % values refer to entire sample.

In order to examine the possibility that there might be some variation of preference patterns with income, we broke down the figures in Table 2.8 by income group to generate Table 2.9. No obvious trend is evident in first preferences, which remain as before. Second preferences are again much as in the aggregate figures, with the exception of the

TABLE 2.9 Ranking of characteristics desired in soap for laundry use, by income group (% of households in each income bracket)*

First rank

Monthly household income (Tk)	Washing ability	Durability	Scent	Wrapping	Resistance to breakage	Colour
100–149		100				
150–199	100					
200–249	67	23				
250–299	100					
300–399	89.6	10.4				
400–499	88	12				
500–749	95.7	4.3				
750–999	91.6	5.6			2.8	
1000–1499	88.4	9.3			2.3	
1500–1999	81.2	12.5			6.3	
2000 and over	94	6				

* Not all households gave second and third rankings; % values refer to entire sample.

Second rank

Monthly household income (Tk)	Washing ability	Durability	Scent	Wrapping	Resistance to breakage	Colour
100–149						
150–159		40				
200–249		30			7.6	
250–299		60				
300–399		65.5				
400–499	12	52				
500–749		40.4			2	
750–999	5.6	61			2.8	
1000–1499	7	67		2.3	7	
1500–1999	6.3	69			12	
2000 and over	6	3	17	43	29	

Table 2.9 cont.

Third rank

Monthly household income (Tk)	Washing ability	Durability	Scent	Wrapping	Resistance to breakage	Colour
100–149						
150–199						
200–249						
250–299						
300–399						
400–499						
500–749			2			
750–999		2.8		2.8		
1000–1499			2.3			
1500–1999						
2000 and over			26	14	29	

responses from the thirty-five members of the highest income bracket. The response ratio here was the highest, at 98 per cent, to the request for a second preference, and no fewer than 60 per cent noted the 'aesthetic' aspects – scent and wrapping – as important. Third preferences were similar to those in Table 2.8. Thus, income difference seems to have little effect on the perceived requirements of soap for laundry purposes, the only exception to this rule being the greater emphasis placed on non-functional properties by the highest income group.

Turning now to soaps used for toilet purposes, we find (see Table 2.8) a rather broader spectrum of requirements. Washing ability was again regarded as the main priority by most respondents, although now by only 52 per cent of all households, while a third of all households regarded scent as the most important attribute. Second preferences were equally divided among washing ability, durability and scent, with resistance to breakage, wrapping, and colour (for the first time) also being mentioned. Third preferences were distributed across all six properties, with wrapping, scent and colour being mentioned by about 24 per cent of all households (or just over half of all respondents). Thus, the expected emphasis on the non-functional attributes of soap used for toilet purposes was apparent although the practical requirement for good washing ability remained the key attribute. The wider spread of responses across attributes reinforces the conclusion that consumers take a less severely practical view of the required properties of a toilet soap than of those of soap for laundry use.

However, this conclusion does not apply to all income groups. In Table 2.10 the rankings are shown by income group, and it will be seen that preferences are closely related to income. Thus the emphasis on washing ability is strongest among the low-income groups, and drops steadily as income rises. The reverse is true with regard to scent; apart from one outlying case, the emphasis on scent as the most important attribute of soap for toilet use increases with income, and no fewer than

TABLE 2.10   Ranking of characteristics desired in toilet soap, by income group (% of households in each income bracket)*

First rank

Monthly household income (Tk)	Washing ability	Durability	Scent	Wrapping	Resistance to breakage	Colour
200–249			100			
250–299		100				
300–399	100					
400–499	80	20				
500–749	81	11.5	7.5			
750–999	57.5	12.0	27.5		3	
1000–1499	55	21.5	24			
1500–1999	46	15	38			
2000 and over	17	6	74	3		

*   Not all households gave second and third rankings; % values refer to entire sample.

Second rank

Monthly household income (Tk)	Washing ability	Durability	Scent	Wrapping	Resistance to breakage	Colour
200–249		100				
250–299					100	
300–399		50	50			
400–499	20	20	10		50	
500–749	19	38.5	19		4	
750–999	27.5	40	18	6	3	
1000–1499	24	21.5	38		5	2.5
1500–1999	15	38	38	7.7	7.7	
2000 and over	51.5	17	14	6	12	

Table 2.10 cont.

Third rank

Monthly household income (Tk)	Washing ability	Durability	Scent	Wrapping	Resistance to breakage	Colour
200–249						
250–299						
300–399						
400–499			40			
500–749	4	4	11.5	4		
750–999	9.5	9.5	21	12	3	
1000–1499	2.5	24	7.5	10	2.5	2.5
1500–1999	15			7.7	7.7	7.7
2000 and over	14	12	3	28.5	6	

74 per cent of respondents in the top-income bracket mentioned it. Taking second- and third-ranked properties into account, it is clear that the demand for non-functional characteristics – scent, wrapping and colour – is positively related to income; whereas that for functional characteristics – washing ability, durability and resistance to breakage – is negatively related to income. This does not mean that households in the low-income group make no distinction between the attributes of soap for laundry and toilet purposes, but they make less of a distinction than do higher-income households.

Matching the demand for soap to demand for properties. We have ascertained how purchases of soap are distributed between income groups, the uses to which it is put, and its desired properties. However, the question remains – do the soaps purchased possess the desired properties?

Over 99 per cent of all respondents earning up to Tk1499 per month used soap mainly for laundry purposes and 86 per cent regarded it as useful for toilet purposes (although this was a secondary consideration). This might imply that the properties demanded in the first instance by this large group of consumers will be those regarded as desirable in soap used for laundry purposes – namely washing ability, durability and (less important) resistance to breakage. Looking now at the soaps actually purchased, we find that demand for washing soaps by the households in the income range under discussion does in fact exceed that for toilet soap. Thus the aggregate monthly demand for labour-intensive washing soap was Tk1103, that for capital-intensive washing soap Tk2219 and that for toilet soap Tk1200. (In the income ranges above Tk1499 the corresponding totals were Tk184, Tk2020 and Tk1292.80.) It is interesting that although the pattern of use shown in Table 2.11 remains essentially unaltered up to an income level of Tk1499, rapid shifts occur in consumption away from the labour-intensive washing soap and towards the two capital-intensive soaps. Somewhere above Tk400 per month the purchase of toilet soap becomes an important means of satisfying the demand for soap for non-washing use; at incomes below this level, washing soap provides the necessary non-washing properties.

To what extent is this shift in purchasing patterns 'rational'? Some subjective judgement may be necessary here, but we can at least be sure that the shift to toilet soap is not dictated by *its* principal quality ('washing ability') as perceived by a majority of purchasers, as washing soaps are equally effective in this regard as they are with respect to durability and resistance to breakage. We may conclude therefore that the progressive expansion of per capita and per household demand for toilet soap at higher-income levels reflects a demand for the 'non-functional' properties of such soap – scent, wrapping and colour – properties that certainly cannot be supplied adequately by either of the two washing soaps. Thus the demand for properties of toilet soap *is* correctly matched by the attributes of the soap purchased. However, the *main* properties demanded are also provided by other cheaper soaps. Since no market exists for products satisfying individual attributes (indeed, this is the crux of the Lancaster argument) it is impossible to isolate and examine the demand for the non-functional attributes of toilet soap. However, it seems reasonable to suppose that as income declines, 'non-functional' demands weaken first, leaving the demand for 'functional' properties intact. These demands can, however, be met by cheaper washing soaps.

The situation with regard to washing soaps is more complex because the same set of attributes is demanded of two similar soaps produced by different technologies. Although it has proved difficult to discern clear relationships underlying the demand for the labour-intensive soap, it is apparent that the shift to capital-intensive washing soaps is related to income levels, with the balance of consumption shifting heavily in favour of the latter as income rises (especially in the urban areas). The key question in the present context is whether the attributes demanded of washing soap are in fact better fulfilled by the more expensive soap variant, in which case the shift towards it as income rises reflects only the

TABLE 2.11 Distribution of consumption by soap type and income class (in %)

Household income (Tk/month)	Washing soap (labour-intensive)			Washing soap (capital-intensive)			Toilet soap (capital-intensive)			Total soap*			Proportion of total consumption for the country as a whole	
	Rural	Urban	Total	Rural	Urban	Total	Rural	Urban	Total	Rural	Urban	Total	By group	Cumulative
0–99	–	–	–	–	–	–	–	–	–	–	–	–	–	–
100–149	100.0	–	100.0	–	–	100.0	–	–	100.0	100.0	–	100.0	0.5	0.5
150–199	100.0	–	100.0	–	–	100.0	–	–	100.0	100.0	–	100.0	3.0	3.5
200–249	76.8	–	76.8	5.9	–	5.9	17.2	–	17.2	100.0	–	100.0	4.8	8.3
250–299	69.2	–	69.2	30.8	–	30.8	–	–	–	100.0	100.0	100.0	5.8	14.1
300–399	78.7	0.0	74.1	19.2	100.0	24.0	2.1	–	1.9	100.0	100.0	100.0	12.9	27.0
400–499	61.1	11.7	57.6	26.6	62.5	29.1	12.3	25.8	13.3	100.0	100.0	100.0	11.7	38.1
500–749	48.3	23.5	46.4	43.7	48.6	45.1	8.0	27.8	9.5	100.0	100.0	100.0	21.8	60.5
750–999	25.3	7.1	23.7	66.2	61.4	65.7	8.5	31.4	10.5	100.0	100.0	100.0	12.6	73.1
1000–1499	52.7	2.6	46.3	7.0	60.4	13.9	40.3	36.8	39.0	100.0	100.0	100.0	17.1	90.2
1500–1999	69.0	3.6	61.1	18.6	54.1	22.9	12.4	42.3	16.0	100.0	100.0	100.0	4.9	95.1
2000 and over	12.6	–	10.9	50.3	62.9	52.0	37.2	37.1	37.1	100.0	100.0	100.0	4.8	99.9
Total			52.4			32.0			15.6			100.0		100.0*

* Error due to rounding.

Source. Revised from Table 2.6 using population weights.

inability of poorer consumers to afford soap with the desired properties. Alternatively, consumers may be under the illusion that the soaps are different although they may in fact be equivalent in practical terms.

We recall that the first preference attribute of soap of this type was washing ability and durability (nominated by 90 per cent and 8 per cent of respondents respectively). How do the two types of washing soap perform in these respects? Our earlier discussion of the technology/ product quality relationship suggests that capital-intensive soaps will have the advantage, because they incorporate superior quality fats, which increase the stability of the lather and at the same time reduce the dissolving rate and increase the firmness of the soap. Other things being equal – although prices in particular are *not* equal – the capital-intensive soap thus satisfies the main requirements of users of washing soap better than its labour-intensive rival. It does *not* seem to be the case that the apparent sophistication of the former product favours it to an unjustifiable degree. On the other hand, it is also true that if they had unrestricted access to all inputs on a regular basis, labour-intensive producers could manufacture a washing soap possessing the desired properties. Such a soap would be of a somewhat lower quality than the capital-intensive soaps currently available, but being less expensive to produce, it would be more attractive to consumers than the labour-intensive washing soap available at present.

Market imperfections and the soap industry. So far we have referred only in passing to the influence of market imperfections on the production and consumption of soap. However, in Bangladesh producers are not in fact free to determine the nature of products they produce (given the limits set by the nature of the technology used); neither are consumers free to buy from a full range of soap products on the basis of accurate information on product characteristics. This has an important bearing on the conclusions to be drawn from our earlier analysis.

Product quality. As most of the raw materials required for the production of good-quality washing soap have to be imported into Bangladesh, access to imports is essential to produce high-quality products. The fragmented nature of the operations of the small-scale, labour-intensive segment of the soap industry in Bangladesh is a serious handicap in this respect, for it cuts off most manufacturers from the essential licence-granting procedures. Given a steady flow of raw materials of appropriate quality, the labour-intensive producers could compete more effectively in terms of quality with their larger, capital-intensive rivals. As is often the case in developing countries, bureaucracy has a significant influence on choice of technology – in this case, favouring sophisticated, large-scale operations.

Marketing and consumption patterns. Marketing has a considerable influence on the purchasing pattern. The labour-intensive soap manufacturers in our survey complained about the strong competition from the larger capital-intensive firms. The former have no properly organised marketing system, often relying on hawkers and street vendors as retail outlets. Some have long-established connections with traders in distant rural areas, but only three spend money on advertising. On the other hand, the capital-intensive factories usually have well-organised marketing departments and distribution networks. Moreover, they spend considerable sums of money on promotion activities, especially advertising on radio and television. The smaller firms cannot rival their rivals' promotional activities at present, and it seems likely (although we have only impressionistic evidence on this) that the pattern of demand is affected.

Reinforcing the advertising referred to above is the superior distribution system of the capital-intensive manufacturers. In contrast, the labour-intensive producers cannot match the distribution facilities of the large firms, and their inability to provide regular supplies to recognised outlets has clearly affected the decisions of an appreciable number of purchasers. In the course of the survey of soap users it was observed that for a high proportion of rural users (35 per cent) the greater availability of labour-intensive soap was the key factor in the purchase decision. In the absence of a well-developed retail distribution system, the 'real' pattern of demand may be masked by the inability of the consumer to locate his 'first preference' soap.

2.3 Income Distribution, Consumption of Soap and Employment

Income distribution and the composition of demand

Our discussion of the relationships between demand and income suggested that most of the demand for labour-intensive washing soap is concentrated in the lower half of the income spectrum, especially in rural areas. Moreover, rising incomes encourage a switch to capital-intensive soaps. Although the unsatisfactory nature of the estimated regression equations for labour-intensive washing soap makes it difficult to generate reliable figures for the composition of demand by product at each income level, we do have sufficient information to use the regression results to provide at least a partial cross-check on the 'by-inspection' analysis.

By combining income distribution data[10] with the consumption data in Table 2.7 and census data on spatial distribution of population, we can generate the figures presented in Table 2.11, which indicate how demand (weighted by population) within each income bracket splits among the three types of soap. The pre-eminence of labour-intensive washing soap at lower incomes is clearly shown, as is the general tendency for demand to shift towards capital-intensive washing soap and toilet soap as income rises. Since the marginal propensity to consume soap is constant over most of the income spectrum, switching Tk1.00 of income from a higher to a lower-income group will tend not to affect total soap consumption, although it does alter the com-position of purchases. The conclusion to be drawn from Table 2.11 is that such a redistribution of income will tend to increase aggregate consumption of labour-intensive soap and reduce consumption of the capital-intensive soaps. Indeed, given the crudeness of this calculation and the probability of sampling error, it is quite possible that, excluding once again the lowest income bracket, progressive redistribution *always* increases demand for labour-intensive washing soap. By the same token, a general increase in incomes may have the effect of reducing the total demand for that soap, although encouraging the consumption of capital-intensive soaps.

Such a process seems to operate in a more clear-cut fashion in urban areas. As the data in Table 2.11 show, the decline in the consumption of labour-intensive soap as income rises is sharper in the urban sample, the value falling to zero in the top income bracket. The expansion of the share of the capital-intensive soap as income rises is consequently more rapid in this sample. Differences in tastes stemming from greater exposure to advertising and other influences conditioning demand, including the more effective retail distribution system, may explain this difference in the patterns of rural and urban purchasing. Our regression analysis, it may be recalled, also identified sharp differences in purchasing behaviour in the rural and urban sectors, and provides us with information on two further explanatory variables – household size and occupation. Since there is no systematic difference across income groups in the first of these we can rule it out.[11] But as urban-dwelling is particularly associated with non-manual work which is in turn positively related to a preference for capital-intensive washing soap, and as the manual work is associated with a preference for labour-intensive washing soap, the 'occupation effect' reinforces the tendencies mentioned earlier.

Thus our analysis strongly suggests that preferences for soaps are not random. On the contrary, there is a clear tendency for low incomes, rural location and manual occupations to be associated with the purchase of labour-intensive soap; and for high incomes, urban location, and non-manual occupations to be associated with the purchase of capital-intensive soaps. Assuming continuing rural–urban migration in most developing countries, an upward drift in per capita incomes, and an increase in the ratio of non-manual to manual jobs, purchasing patterns are likely to move towards more capital-intensive soap products. However, since income largely determines demand patterns, redistribution could halt or reverse this trend.

Although the figures for the income groups in Table 2.11 are not very reliable because of the small size of sample, they do indicate the *general* nature of the income–consumption relationship. It seems that the shifting of income from the ranges above Tk750 (which account for around 35 per cent of income but only 15 per cent of households) to lower ranges would strongly favour the consumption of labour-intensive products. This effect would probably be particularly pronounced with respect to redistribution towards the Tk150–399 bracket, where very little capital-intensive soap is purchased by the 43 per cent of the population covered. As noted earlier, the effect of shifting income to the *very* lowest brackets (Tk0–149) might be to reduce over-all soap consumption, since their marginal propensity to consume soap appears to lie below the norm. However, since these income groups include only 7 per cent of all households this effect would almost certainly be swamped by the positive effects of redistribution towards the intermediate income range. Finally, marginal increases in rural incomes at the expense of urban ones might also be expected to shift demand towards labour-intensive soaps.

Employment implications

We have already noted that redistribution is unlikely to affect total demand for the product, but may increase the demand for labour-intensive soap. How many new jobs would be created by such a shift depends on the nature and extent of redistribution, and the degree of under-utilisation of capacity prior to the change.

On the demand side, we know that at most income levels the impact of a change of Tk1.00 in income is to change the demand for soap by between Tk0.01 and Tk0.03 – that is, 1–3 per cent of any absolute change in income. If such a redistribution involved a reduction in the income of households in income groups above Tk399 per month, then almost all the impact on the soap market would take the form of a switch from capital-intensive soaps – for which demand would fall to low levels – to labour-intensive soaps. What would happen to employment in the industry? Given no excess capacity, aggregate installed capacity would be inadequate since the quantity purchased by Tk1.00 spent on labour-intensive soaps is greater than for capital-intensive soaps (about 10–80 per cent more at current prices) and much greater than that for toilet soap (about 300–400 per cent more). Consequently, although the production of soap similar to what we have called labour-intensive washing soap is quite possible in capital-intensive factories, the latter would not have sufficient capacity to satisfy demand. Moreover, in view of the somewhat lower revenue per ton generated by production of lower-grade soap, there must be some doubt as to whether capital-intensive technology is competitive with labour-intensive technology at the prices currently prevailing in product and factor markets in Bangladesh.

Very roughly, then, a shift of Tk1 million of income from the over-Tk399/month (household) income earners to those below that level would have the effect of switching Tk10 000–Tk30 000 worth of demand for soap from one income bracket to another. Of this some 80 per cent (Tk8000–Tk24 000) would shift from capital-intensive to labour-intensive type soap. This could either be produced partly in the capital-intensive sector (which now has some spare capacity) and partly in the labour-intensive sector (which would require to create new capacity) or entirely in the latter if such soap could not be produced competitively using capital-intensive technology. Either way new jobs would be created in the low-technology sector, the number running at 0.3–1.0 new jobs per Tk10 million redistributed[12] (i.e. per 0.02 per cent of GNP redistributed). This implies that any realistic level of redistribution is likely to create only a modest number of new jobs in the soap industry, although this is partly a result of the small size of the industry in Bangladesh.[13]

Clearly, this is a very rough estimate not only because of the inaccuracy inevitably associated with the small sample sizes on which the analysis is based, but also because it depends on the degree of excess capacity available in the labour-intensive sector of the soap industry at the time redistribution takes place, which is substantial in Bangladesh at present; and on the choice of technology and efficiency of operation of that sector, in which inefficient plants use up to twenty times more labour per unit of output than efficient ones.

Further implications of income redistribution

In the previous sections attention was focused on the employment effects of a redistribution of income in favour of the lower-income groups in Bangladesh. However, there are additional secondary effects of some significance. In particular, the expanded use of labour-intensive methods of soap production would increase the demand for locally produced raw materials and plant and machinery. This would concentrate the beneficial spread effects of industrialisation (broadly construed to include conventional 'multiplier' effects and familiarisation with new techniques) in the home economy, and avoid the creation of further pressure on the balance of trade.

Interrelated factors

Household size, occupation and location are known to exert an independent influence on the magnitude and structure of demand for soap. We also know that per capita incomes tend to be higher in non-manual occupations, in larger households, and in urban areas. It may be that a redistribution of income in favour of the lower-income groups would lead to a decline over time in average household size, and hence in economies of scale in consumption: thus, although population might remain unchanged (implying an increase in the number of households) per capita soap consumption would rise, and much of this rise would reflect the increasing demand for labour-intensive soaps. However, apart from implying waste, by deliberately foregoing scale economies, this link between income distribution and consumption/employment is a tenuous one.

Future demand and its employment implications

If per capita incomes, the spatial distribution of population, occupation patterns, income distribution and household size all remained un-changed over the years to 1990, the demand for soap would still rise substantially because of the sharp increase in population.[14] Given our population projections, aggregate demand for soap, and demand for each category of soap, would rise by 34.5 per cent (median estimate).

The implication of an across-the-board rise in per capita incomes is, as was noted earlier, an increase of Tk0.01–0.03 in total demand for soap for every Tk1.00 increase in income. Disaggregating by type of soap, superimposing this effect on the impact on demand of the projected population increase, and making a small upward adjustment to allow for increased urbanisation and a possible switch away from manual occupations, yields projections of the increase in demand for each type of soap. Roughly speaking, demand for each type will rise by between 27 per cent and 42 per cent, the precise increase depending on the rate of growth of population assumed (2–3 per cent per annum).

Given the excess capacity in the Bangladesh soap industry, it is likely that at least that part of the projected expansion of demand for capital-intensive soaps could be supplied using existing facilities. The position is rather different for labour-intensive washing soap, for which new capacity would be needed. The number of new jobs created in this sector could vary considerably, depending on the scale of production facilities used to produce the additional output. The detailed discussion of the employment-generating capacity of alternative technologies in section 2.3 above suggests that the median projected increase in demand for washing soap produced by labour-intensive techniques could be satisfied by setting up new establishments providing between 350 and 5000 new jobs, most of which would be manual. Thus the potential impact of population change on employment in the soap industry in Bangladesh is much greater than the effect of shifts in income distribution of the kind discussed earlier in this section.

Alternative policies to manipulate demand and supply

In the discussion of the attributes of various soaps it was noted that the demand for attributes seems fairly consistent with the observed pattern of consumption. This suggests that consumers do not err in their choice of type of soap, but indeed purchase that which best satisfies their perceived needs. At the same time distinct differences between urban and rural demand patterns were noted, and we have seen that to *some* extent the consumption of labour-intensive soaps is reduced by their intermittent availability, caused partly by the non-availability of imported materials, although the lack of proper marketing channels is also important. It was also noted that the quality of washing soaps made by labour-intensive means would be closer to that of capital-intensive soap if better quality imported materials were used.

In these circumstances an attempt to create employment by draconian measures such as barring the scale of capital-intensive toilet and washing soaps would be very costly. True, most of the properties desired in washing soap and some of those desired in toilet soap *can* be provided by labour-intensive washing soap, but consumers clearly regard the latter as technically inferior. A more efficient and acceptable approach to the manipulation of demand to favour the labour-intensive product might concentrate on strengthening the weak points in its marketing and its technical characteristics. Specifically, more emphasis could be placed on regular supplies to consumers by the establishment of reliable distribution procedures. A modicum of advertising designed to inform the potential customer of the regular availability of this soap (so ensuring regular demand) would help, and might also aid the recapture of urban markets lost to the more sophisticated and expensive soaps.

Again, bureaucratic constraints on the smooth flow of high-quality imported raw materials obstruct regular production and could be removed. The labour-intensive soap producers might be encouraged to import directly; and where smallness or remoteness is an obstacle, the development of an efficient, competitive wholesaling system could be encouraged by the provision of detailed market information to all producers. Given the improvement in the quality of labour-intensive washing soap resulting from the use of good-quality materials, a switch by a part of this sector to higher-quality soap could make considerable inroads into the market for capital-intensive washing soaps. Indeed, the potential here for job creation may be rather greater than that of income redistribution.

On the technology front, the dissemination of information on techniques currently in use in Bangladesh, with cost and profit profiles, might reduce the variation in production methods and encourage efficiency within the labour-intensive sector. A fairly modest local programme could have a significant impact – more than could be expected from a massive externally financed and staffed research project on the industry, and would have the advantages of being cheap, and of producing results quickly.

These various possibilities would of course require some financial support, perhaps in the form of improved credit facilities for labour-intensive producers.

On the demand side, our regression analysis suggests that the occupation of the head of household and the location (rural/urban) of residence influence patterns of soap consumption. Clearly, these variables involve much more far-reaching issues than those currently under consideration, but at least it can be said that policies designed to reduce migration to the towns *may* have, as one side-effect, a favourable impact on the employment situation via their effect on the product structure of demand.

2.4 Concluding remarks

This study was aimed to test the 'income distribution–technology–employment' hypothesis with respect to a product satisfying a basic need – soap. Field investigations on the consumption and production of soap in Bangladesh have enabled us to establish that a range of product quality exists; that quality (and price) are positively correlated with the capital-intensity of technology: and that household (or per capita) income is positively correlated with quality of product purchased. Consequently, income distribution and employment generation are related, via the product quality/technology link, in such a way that a more even distribution of income tends to be associated with a higher level of job creation in this industry.

Pushing the analysis further in order to examine the matching of product characteristics, it was shown that soaps made by labour-intensive technologies appear to be adequate to the task of satisfying consumers' basic needs with regard to soap, although not possessing clearly redundant characteristics. Of the two capital-intensive soaps examined, the toilet soap possesses properties well in excess of the basic needs minimum, whereas the other, a washing soap, has few superfluous properties as such, but appears to satisfy basic needs rather more than is necessary. Little evidence was found of inefficiency in buying behaviour stemming from lack of information, although it did seem that advertising and the superior marketing of capital-intensive soaps do affect those consumers exposed to them. Moreover, many cases were found where the failure of supply of the labour-intensive product forced consumers to purchase capital-intensive soaps.

As regards the choice of technology, no clear differences emerged in the respective rates of return of technologies with widely differing factor proportions, although there was some evidence that economies of scale may be more powerful at the capital-intensive end of the spectrum, and several clearly inefficient labour-intensive technologies were identified. It also proved difficult to discern systematic differences in skill requirements although the difference in the unskilled labour input per unit of output between capital-intensive and labour-intensive technologies was considerable. In addition labour-intensive methods were found to be unsuited to the production of high-quality soap, although they have the advantage of drawing more of their capital and current inputs from the domestic economy.

Tentative estimates of the primary impact on employment of an income redistribution in favour of the low-income section of the population suggests that the effect would be fairly modest, especially where there is already excess capacity, as in Bangladesh at present. However, the impact of redistribution would be considerably enhanced if efforts were made to improve the quality of the labour-intensive product by ensuring regular supplies of high-quality imported raw materials, and to reorganise the marketing of the product rather than rely on the present somewhat erratic system. These two measures could have a considerable impact on the share of the market for washing soap captured by labour-intensive producers, especially if reinforced by a programme of information dissemination on more efficient methods of labour-intensive production.

Appendix: Appropriate Products, Employment and Income Distribution in the Soap Industry: The Case of Ghana

Introduction

A parallel study to that reported in the main body of this paper was carried out in Ghana.[15] The methodology used was as for the case of Bangladesh, though the samples of producers and consumers were somewhat smaller. In general the results of the Ghana study confirmed those derived for Bangladesh, the differences tending to be matters of emphasis rather than kind. A brief summary of the main points is given below, the parallels with the Bangladesh study being noted in each case.

Principal findings of the Ghana study

1. Income distribution in Ghana displays the marked inequality typical of LDCs at the country's level of development; unemployment is also a severe and growing problem.

2. As in Bangladesh, the soap industry in Ghana divides clearly into two sectors – a modern, large-scale, capital-intensive sector, drawing most of its equipment and raw materials from abroad, and a locally oriented, small-scale, labour-intensive sector. Once again, a number of

'technically efficient' technologies were identified, spanning a wide range of factor proportions. And again the capital-intensive methods yielded the higher NPVs, although various 'external' and 'social' factors not included in the DCF calculation might go a long way towards offsetting this difference.

3. As regards consumption, the same three basic soap types identified as important in Bangladesh were again central in Ghana; labour-intensive washing soap was purchased principally by the lower-income earners, and as household incomes rose, so consumption of capital-intensive washing soap and capital-intensive toilet soap also rose. Although the over-all income–consumption relation was similar in the two countries, labour-intensive washing soap was consumed in relatively smaller quantities in Ghana, and the distinction between rural and urban consumption patterns was much less clear-cut. Both these differences may reflect Ghana's higher standard of living, and may indicate the likely path of development of the Bangladesh industry in the absence of countervailing action – the rapid elimination of the labour-intensive sectors and the extension of 'modern', urban consumption patterns to the rural areas. Certainly, in neither country have the small, labour-intensive producers been able to combat effectively the superior advertising and marketing methods of the large-scale, capital-intensive sector; the combination of irregular access to inputs (and hence irregular delivery of final product) and lack of a brand name were as damaging in Ghana as in Bangladesh.

With regard to the relationship between the demand for properties of soap and the actual pattern of soap purchases, the findings of the Ghana study again mirrored the Bangladesh findings. Toilet soap was found to be bought by a number of members of the sample of consumers when the properties demanded were, in fact, those possessed by washing soap – and to this extent demand was wrongly 'targeted'. With regard to washing soap, here the tendency to prefer capital-intensive washing soap despite its higher price was found to reflect a genuine appreciation of the superior properties of the latter; consumers preferred the capital-intensive variety, and bought it when they were able to afford it. (However, with some improvement in the quality of raw material inputs, labour-intensive washing soap would be much more competitive here.)

As before, multiple regression analysis was used to model the consumption relationship. On the whole, the results for Ghana were similar to those for Bangladesh, with per capita income emerging as the dominant regressor and household size as the next most important regressor. However, as noted earlier, the rural–urban distinction that was striking in Bangladesh was not found in the Ghana study in the cases of the individual soap types.

4. Finally, the over-all conclusions of the Ghana study were that static income redistribution would at best provide only a very small increment to employment opportunities, and that future growth in population and income were likely to be very much more significant factors in increasing employment in soap production in the future.

These conclusions are very similar to those of the Bangladesh study. They are perhaps best interpreted as a powerful confirmation of the finding that appreciable numbers of new jobs can be created in the soap industry. However, as was found to be the case in Bangladesh, the additional employment is more likely to be generated by encouraging the regeneration of the small-scale, labour-intensive sector in order to enable it to take advantage of the rapid market growth which seems likely in the future, rather than by any form of income redistribution.

Notes

1. A. K. A. Mubin collected the basic data for this study while a postgraduate student at the University of Strathclyde (UK), where he was assisted by David Forsyth, Reader in the Department of Economics.
2. These are summarised in A. Mubin and D. Forsyth, 'Appropriate Products, Employment and Income Distribution in Bangladesh: A Case Study of the Soap Industry' (Geneva: ILO, 1980; mimeographed World Employment Programme research working paper; restricted).
3. However, for some comments on their future use see ibid, appendix D.
4. Saponification is a chemical reaction between fats (e.g. tallow) and an alkaline solution (most commonly, caustic soda).
5. Tallow, liquid caustic soda, water and steam are required here. The raw materials are pumped into the saponification pan. Soap and glycerine are produced.
6. The questionnaire on which these discussions were based is reproduced in Mubin and Forsyth, *Appropriate Products, Employment and Income Distribution in Bangladesh.*
7. All data quoted in the text or in tables are drawn from our field study unless otherwise indicated.
8. This is reproduced in Mubin and Forsyth, *Appropriate Products, Employment and Income Distribution in Bangladesh,* appendix C.
9. For a more elaborate discussion, see ibid, pp. 61–72.
10. This is presented in ibid, appendix table 2.
11. Ibid p. 95 and appendix 3.
12. This figure is based on an income elasticity of demand for all soaps of between 1 and 3; and the expectation that the shifting of Tk 10 m of aggregate

demand through a redistribution of income towards the lower-income brackets will lead to a shift of between Tk80 000 and Tk240 000 demand from capital-intensive to labour-intensive soaps. With a price of Tk4 per pound, and 82.2 pounds per maund, this implies an additional demand of between 243–730 maunds. The employment impact is computed on the basis of the labour input per unit of output in the most efficient labour-intensive factories mentioned earlier in this chapter. The figure turns out to be between 0.6 and 2.0, although this is partly offset by the loss of jobs in the capital-intensive sector, thus leading to the final estimate of a net employment impact of between 0.3 and 1.0 (i.e. 30–100 per 1 per cent of GNP).

13. It is to be expected that some other industries will display a similar income distribution–employment relationship to that observed here, so that the *total* elasticity of employment with respect to changes in income distribution will be considerably greater than that for the case of soap alone.

14. The reasoning behind this statement is elaborated in Mubin and Forsyth, *Appropriate Products, Employment and Income Distribution in Bangladesh*, pp. 102–4.

15. The full text of the report from which this Appendix is drawn is available in mimeo form as: A. K. A. Mubin and David J. C. Forsyth, 'Appropriate Products, Employment and Income Distribution in Ghana: A Case Study of the Soap Industry'.

3 Basic Needs and Appropriate Technology in the Malaysian Bicycle Industry

Fong Chan Onn[1]

3.1 Introduction

This study on the bicycle, the simplest technique of personal transport other than walking, and therefore a basic need in many developing countries, is based on two sets of data. The first concerns consumers' perception of the bicycle as a product (i.e. its basic and non-basic characteristics) collected in a survey of consumers. The second set of data concerns the production techniques applied by bicycle manufacturers. These data were obtained in interviews with various firms. To collect the first set of data it was necessary to conduct a mini-household survey of about 200 households stratified along the urban–rural strata in the ratio 40:60. The information gathered from each household included data on average household expenditure and the ownership of household assets, and on consumer perceptions of the brand attributes of bicycles and their purchasing behaviour *vis-à-vis* bicycles.

In Malaysia, the two major bicycle manufacturers are Raleigh (Malaysia) and Far East Metal Works. The latter produces a variety of non-Raleigh brands sold throughout the country. Apart from the two major manufacturers (to be referred to here as the primary firms), about eleven other establishments manufacture some parts of components of the bicycle (ancillary firms) for sale to the two primary firms and other bicycle retailers. In this study, the two primary firms and a number of ancillary firms were interviewed. The information collected includes: (i) general background information, (ii) fixed capital, (iii) employment and payroll, (iv) technology, (v) marketing, and (vi) problems and prospects.

The data collected were analysed using conventional statistical and economic methods. These are described in the subsequent sections. This chapter has the following structure. In section 3.2, an overview of the bicycle industry in Malaysia is given. This is followed by a discussion of the consumer's perception of the bicycle as a product in section 3.3. In section 3.4, the data collected on the manufacturing technologies are analysed and the technology that is closest to being an appropriate technology is discussed. The relationships between income distribution, the application of appropriate technology and employment generation are discussed in section 3.5, which also presents some conclusions about the role of government policy.

3.2 The Development of the Malaysian Bicycle Industry

Components of a typical bicycle

There are several types of bicycles, of which the roadster is the most common. The components required in a typical roadster are set out in Table 3.1, which shows that the frame, fork, mudguard, handlebar, chain guard, luggage carrier and stand require raw material inputs consisting mainly of steel tubes and sheets; and that relatively simple processes of cutting, grinding, tread forming, pressing and bending, and welding suffice for their manufacture. Most well-equipped workshops are capable of performing these operations although, because of economies of scale, workshops specially set up for the purpose can make the components at a more consistent quality and lower unit cost. The other parts require a higher level of technology involving more sophisticated machines working on semi-finished inputs like high carbon steel. They are generally produced by specialised firms and the minimum economic plant size is between 50 000 and 150 000 units a year.

Large manufacturers

In spite of its relatively simple technology, almost all bicycles were imported until the mid-1950s.[2] In 1959, the Far East Metal Works Company was formed in Petaling Jaya with an initial capital of $0.6 million to manufacture certain components and assemble bicycles. At present, the company manufactures bicycles under the brand names 'Seven-Up', 'Rolex', 'Apache', 'Yokohama' and 'Diamond'.

The Raleigh Cycle (Malaysia) Bhd, formed in 1967 by Raleigh

TABLE 3.1   Main components of a typical roadster bicycle

Component	Raw material requirements	Manufacturing processes
1 Body frame	Steel tube and joints	1 Tube cutting 2 Grinding and deburning 3 Tread forming 4 Pressing, reaming and bending 5 Welding and brazing 6 Polishing and painting
2 Fork	Steel tube	As for body frame
3 Mudguard	Steel sheet	As for body frame but sheet cutting instead of tube cutting
4 Crank, chain wheel and chain	High carbon steel	Specialised machines*
5 Handlebars	Steel tube, steel level and rubber handle grip	As for frame together with electroplating (nickel and chromium plating)
6 Brakes	Steel lever, brake shoes	Specialised machines*
7 Hubs, spokes and nipples and rims	High carbon steel sheet and wire	Specialised machines*
8 Saddles	Steel wire for springs, leather	Specialised machines*
9 Pedals	Dust caps, and pyramid full rubber	Specialised machines*
10 Chain guard	Steel sheets	As for mudguard
11 Luggage carriers	Steel tubes and sheets	As for mudguard
12 Stand	Steel tubes and sheets	As for mudguard
13 Bell	High carbon steel	Specialised machines*
14 Tyre and inner tubes	Rubber, canvas, and steel wire	Specialised machines*
15 Other accessories e.g. lock, lamp, etc.	Miscellaneous	Specialised machines*

* For details of the machineries and processes, see United Nations, *Bicycles: A Case Study of Indian Experience* (Vienna: United Nations Industrial Development Organisation, 1969).

Industries Ltd of the United Kingdom to manufacture and assemble the world-famous Raleigh bicycle, represented a very major step in the transfer of bicycle technology to Malaysia. The company was formed with a paid-up capital of $2.15 million, 50 per cent Raleigh equity participation with 'pioneer status' but with no tariff concessions on its imported components. Its brands are 'Hercules', 'Robin Hood' and 'Raleigh'.

Apart from Raleigh and Far East, no other large bicycle manufacturers are presently operating in Malaysia and marketing bicycles under brands with which the establishment is identified.

Bicycle ancillary firms

In Malaysia, a number of establishments are manufacturing some parts or components of a bicycle. Although no complete listing of these establishments exists, attempts to identify them through the *Directory of Manufacturing Establishments*, the Yellow Pages in the telephone directory and the bicycle dealer associations revealed eleven establishments spread out evenly over the country. These firms manufacture all kinds of components with the exception of cranks and chain wheels, hubs, pedals and bells. Besides low-technology items like body frames, forks, mudguards, handlebars, chain covers, luggage carriers and bicycle stands, relatively high-technology items like chains, brake system spokes, nipples and rims are manufactured. Firms that manufacture the latter items were all established after 1970, enjoying some form of fiscal incentives from the government.

Tariff structure

Since the formation of Malaysia in 1963, the tariff levied on each imported bicycle (completed) has been increased from $18 to the current level of $60.[3] The tariff levied on an imported component varies from, for example, $20 on a body frame to $1 on parts of brake systems. The total tariff on all the parts (levied separately) of a bicycle amounts to $74.

The aim of a system of tariffs on bicycle components ought to be to offer greater protection for more sophisticated components (e.g. hubs, crank and chain wheel) so as to encourage their local manufacture, and less protection to components for which manufacture is fairly well established (e.g. tyres, frames, etc.) so as to increase the efficiency of these establishments leading them to look to the export market. However, the tariff structure announced in 1978 offers uniformly high protection on all components, for example, $20 for a body frame and $15 for a luggage carrier. Although in the short run this may benefit local bicycle manufacturers, in the long run it is unlikely to stimulate them to be export-oriented even in respect of those components in which local firms have a strong international comparative advantage because of factor endowments.

Production and employment

Until 1969 the domestic bicycle industry remained a relatively small sector – generating a value added at market prices of only about $0.4 million and employment of less than 200 people. However, the opening of Raleigh (Malaysia) increased value added to about $1.6 million in 1969 and $2.4 million in 1972, that is, an increase of about $2.0 million per annum in three years. Employment also increased to about 390 in 1969 and 500 in 1972. Labour productivity varied from $2600 to $2900 per worker up to 1967; after 1967 it increased to between $4000 and $5400 per worker. This is significantly below the average labour productivity of $5179 per worker in the transport and communications sector.[4]

Market demand and market share

In Malaysia the ownership of bicycles requires no official registration. Furthermore, although *Malaysian External Trade* provides data on the number and value of bicycles traded, the *Survey of Manufacturing Industries* does not give the number of bicycles produced domestically. Thus, any estimate of the domestic market for bicycles must be based on its relationship to the economic environment and past sales. The Malaysian Bicycle Dealers Association's estimate of the domestic market for bicycles is presently about 200 000 per annum.

The minimum economic size for a bicycle plant manufacturing simple pressed and bent parts and purchasing other components from ancillary firms is about 25 000 units per annum. The domestic market is more than sufficient for the two large bicycle manufacturers currently operating, and there appears to be room for about two more bicycle plants. The minimum economic size for a specialised plant producing high-technology components like cranks, chain wheels, hubs and pedals is about 100 000 units per annum. Given the present size of the domestic market, even with a market share of about half the total, ancillary firms producing these specialised components can be viably established in the country.

The bicycle marketing network

Both Raleigh (Malaysia) and Far East market their bicycles in 'completely knocked down' packs to bicycle dealers throughout the country.[5] These dealers in turn sell the bicycles to the retailers (bicycle shops) either in assembled form or in knocked-down packs to be assembled by the retailers. In general, the Raleigh brands retail in the range $200–240 per unit, the Far East brands in the range $150–180, while local brands (assembled by bicycle retailers from purchased components) sell for $100–150. Since the price for a bicycle is relatively low, most individual purchases are on a cash basis. The bicycle ancillary firms also market their components and parts produced through the formal marketing system.[6] They sell their components to the bicycle dealers, who in turn sell them to retailers for both the new and replacement market.

The marketing system for bicycles is relatively efficient. Being small, with an extensive network of roads and railway lines, bicycles produced by the manufacturers in any part of the country can be shipped to another part in a few days. For this reason the prices of bicycles are fairly uniform throughout the country.

3.3 Consumer Perceptions of the Bicycle

In this section we present the consumer's perception of the bicycle as indicated in a survey of 195 bicycle-owning households spread throughout Peninsular Malaysia. The results of this survey are important because consumer perceptions and purchase patterns are among the important inputs that should be considered when determining the most appropriate technology for bicycle manufacturing.

Consumer survey

The consumer survey conducted in 1978 aimed to determine the expenditure pattern of each consuming household, its preference in utility attributes for a bicycle, its perception of brand attributes and its past bicycle purchase pattern. The questionnaire for the survey was refined and adapted from marketing questionnaires designed for similar purposes.[7] The sample was stratified to reflect accurately the conditions in the country. The number of households (rural and urban) included for each State was determined on the basis of its 1970 population[8] relative to that of Peninsular Malaysia as a whole. An analysis of certain important socioeconomic characteristics of the sample (e.g. the mean size and the mean occupational level of the households) confirmed that the sample was fairly representative of the national conditions.

Crucial attributes affecting consumer purchases

In the consumer survey, each consuming household was offered a total of seventeen bicycle attributes[9] and was asked to sort each characteristic into the four attribute categories of: (i) crucially important, (ii) highly important, (iii) fairly important, and (iv) not important to choice. Table 3.2 analyses the consumer's perception of the attributes crucial to choice of a bicycle. From the table it may be seen that the four attributes believed to be crucial to the choice of bicycle are 'reliability as a means of transport', 'low maintenance cost', 'low purchase price' and 'ease of getting compatible parts for repair'. In terms of mean score,[10] each attribute has a score of between 3.67 (for 'reliability as a means of transport') and 3.06 for 'low purchase price'). Thus, in an over-all sense, each of these four attributes is somewhere between crucially important and highly important to choice.

A principal component factor analysis performed on the seventeen attributes showed that the attributes can be meaningfully partitioned into six underlying dimensions of (i) economy, (ii) prestige, (iii) reliability and convenience, (iv) design and workmanship, (v) specialised preference, and (vi) resale value. These six dimensions together account for 60 per cent of the variance in attribute ranking and can therefore be considered as fairly adequate in terms of characterisation of the attributes. Table 3.2 shows that the two most important dimensions with respect to choice of bicycle are 'economy' and 'reliability and convenience'. In terms of mean score, the economy factor has a mean consumer rating of 3.28 while the reliability factor has a mean consumer rating of 3.16 indicating that, on an over-all basis, both these dimensions are rated between crucially and highly important. The other four factors each have a mean rating of less than 3 – or less than highly important.

The economy dimension includes both the low purchase price and low maintenance cost attributes. These two attributes are non-complementary in the sense that a brand with low purchase price tends to be of lower quality and hence requires higher maintenance cost. It is interesting to note that a higher percentage of consumers rate the latter attribute than the former as crucial. This implies a subjective trade-off in the mind of the consumer between low purchase price and low maintenance cost to arrive at a model that is most economic. The reliability and convenience dimension includes the attributes of reliability, ease of getting compatible parts for repair, and ease of getting the machine repaired when it breaks down. These attributes are complementary and together indicate the dependability of the bicycle as a

TABLE 3.2 Consumer preference on attributes crucial to choice of bicycles

Attribute	Attribute dimension	% Consumer who feel attribute is crucial	Mean rating of attribute*	%Consumer who feel attribute dimension is crucial†	Mean rating of attribute dimension†
1 Low purchase price	Economy	46.2	3.06	54.9	3.28
2 Low maintenance cost		63.6	3.49		
3 Prestige as a household item	Prestige	9.7	1.85	6.2	1.66
4 Status symbol for family		2.6	1.46		
5 Reliability as a means of transport	Reliability and convenience	77.4	3.67	45.1	3.16
6 Ease of getting compatible parts for repair		39.5	3.15		
7 Ease of getting machine repaired		18.5	2.67		
8 Robustness of machine – can ferry heavy goods		28.7	2.94		
9 High gearing ratio	Design and workmanship	32.8	3.06	24.7	2.80
10 Good paint finish		19.5	2.57		
11 Good engineering and assembly work		31.8	2.97		
12 Abundance of compatible accessories		10.8	2.45		
13 Bicycle is a good lady-design	Specialised preference	4.6	1.78	4.3	1.67
14 Bicycle is a good sports design		5.6	1.61		
15 Bicycle is a good children's design		2.6	1.62		
16 High resale value	Resale value	21.5	2.42	16.2	2.31
17 Ease of trading-in bicycle when buying new bicycle					

* Computed on a basis of following weights – crucially important (4), highly important (3), important (2), and somewhat important (1).
† Computed on basis of average for all attributes in the dimension.

means of transport. In the next section we shall further analyse the consumer's perception of these two important attribute dimensions.

Income effect on crucial attribute dimensions

A consumer's perception of important attributes in a product obviously is a function of his own income. Household-income data are extremely difficult to gather accurately by a survey. Besides the obviously sensitive nature of the question, which is unlikely to meet with forthright or accurate responses, another important factor is the ambiguity of the household-income concept.[11] For these reasons, this study uses household expenditure as a surrogate for income.

Tables 3.3 and 3.4 illustrate the expenditure effect of the two

TABLE 3.3 Household expenditure with economy dimension

Household expenditure ($/month)	% of household rating*				Mean rating of group	Total households in group
	Crucially important (1)	Highly important (2)	Important (3)	Somewhat important (4)		
Less than $300	65.0	18.0	10.0	7.0	3.41	93
$301 to $500	53.0	21.7	15.4	9.9	3.18	72
$501 to $700	28.8	41.9	22.8	6.5	2.93	19
More than $700	28.3	41.2	21.3	9.3	2.88	11
Total	54.9	23.0	13.9	8.2	3.28	195

* Rating for economy dimension is computed on basis of average for all attributes in the dimension. $X = 17.0$, which means that the hypothesis that the two variables are independent can be rejected at 0.95 level of confidence.

TABLE 3.4 Household expenditure with reliability dimension

Household expenditure ($/month)	% of household rating*				Mean rating of group	Total households in group
	Crucially important (1)	Highly important (2)	Important (3)	Somewhat important (4)		
Less than $300	47.8	25.1	16.7	10.4	3.10	93
$301 to $500	47.4	32.7	9.2	11.7	3.13	72
$501 to $700	55.1	30.5	8.8	5.6	3.35	19
More than $700	52.7	23.1	15.8	8.4	3.20	11
Total	48.6	27.3	13.1	10.5	3.16	195

* Rating for functional dimension is computed on basis of average for all attributes in the dimension.
$X^2 = 2.95$, which means that the hypothesis that the two variables are independent can be accepted at 0.95 level of confidence.

dimensions crucial to the choice of a bicycle. An examination of Table 3.3 shows that there is an expenditure effect on the economy dimension. In general, the smaller the total household expenditure, the more crucial is the economy dimension. In households with an average expenditure of less than $300 per month, over 65 per cent felt that economy is crucial to choice; this category reduces to 28 per cent for households with more than $700 per month. In terms of mean group rating, Table 3.3 shows that households with expenditures of less than $300 per month gave a mean rating of 3.41 (midway between crucially and highly important) to the economy factor. The group mean rating is seen to decrease with the group household expenditure: at over $700 per month, the mean rating is 2.88 (or less than highly important). On the other hand, Table 3.4 indicates a distinct absence of expenditure effect on the reliability dimension. For households of all expenditure categories, about 50 per cent of households feel that the reliability dimension is crucial. The mean group rating for the reliability factor is about 3.20 for each expenditure group.

In summary, then, the reliability dimension is crucial in the choice of bicycle in households at all levels of income. However, there is a distinct income effect on the economy dimension, with most households in the lower-income categories (and fewer percentages of the higher-income categories) rating it as crucial.

Brand groups and crucial attributes

Bicycles in Malaysia can generally be classified into four brand groups: (A) Raleigh brands (e.g. 'Raleigh', 'Hercules' and 'Robin Hood'); (B) non-Raleigh brands manufactured by Far East (e.g. 'Yokohama', 'Apache', 'Grifin', 'Rolex', etc.); (C) non-Raleigh brands assembled by bicycle retailers using mainly imported components (e.g. 'Reddy', 'Oryx', 'Rollick', etc.); and (D) non-Raleigh brands assembled by bicycle retailers using mainly components produced locally (e.g. 'Raffles', 'Butterfly', 'Golden Dragon', etc.).

This section discusses the consumer's perception of bicycles in these brand groups with respect to the attributes identified as crucial to choice.

For each attribute, each consumer was asked to rank the four brand groups of bicycles. For example, for low purchase price a consumer may rank (D) (B) (C) (A) in ascending order of purchase price. These rankings were translated into numerical scores by assigning a weight of 4 to the best brand group, 3 to the second best, and so on. Table 3.5 gives the consumer group mean numerical rating of the four brand groups for

TABLE 3.5 Brand groups and crucial attributes

Attribute	Consumer mean rating of brand group for attribute* (A)	(B)	(C)	(D)	Attribute factor	Consumer rating of brand group for factor† (A)	(B)	(C)	(D)
Low purchase price	1.22	2.88	2.48	3.46	} Economy	2.49	2.62	2.44	2.54
Low maintenance cost	3.56	2.37	2.44	1.63					
Reliability as transport	3.84	2.17	2.59	1.45	} Reliability and convenience	2.85	2.57	2.21	2.53
Ease of getting compatible parts	2.22	2.73	1.99	3.06					
Ease of getting machine repaired	2.48	2.82	2.04	3.07					

* Computed on basis of following weights for each consumer rating:
(4) for best, (3) for second best, (2) for third best, and (1) for last brand group.
† Computed on basis of average rating for all attributes in dimension.

the five attributes crucial to choice. (The mean ratings vary between 1 to 4, in ascending order of strength.)

An examination of Table 3.5 shows that it may be seen that for the economy factor, Raleigh brands are ranked very high on 'low maintenance cost' but very low on 'low purchase price'. The retailers' brands using locally produced components, on the other hand, are ranked very low on 'low maintenance cost' but high on 'low purchase price'. The position of the brands with respect to the two attributes of the economy factor is shown vividly in Figure 3.1, from which it is apparent that consumers view groups (A) and (D) as opposites – (A) has a high price but requires little maintenance, while (D) is the reverse. Groups (B) and (C) occupy rather central positions, that is, they have a lower purchase price but higher maintenance costs than (A) but higher purchase price and lower maintenance cost than (D).

For the attributes associated with the reliability and convenience factor, we can without loss of information restrict our discussion of brand positions in the reliability and convenience factor space to the two attributes of 'reliability as a transport mode' and 'ease of getting compatible parts'. Figure 3.2 makes it clear that group (A) is rated as very reliable but some difficulty is perceived in obtaining compatible parts. Group (C) is perceived as inferior to (A) in both attributes. Group

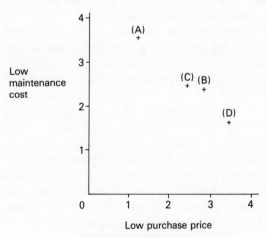

Note: The numbers on the axis are mean consumer ratings.

FIGURE 3.1 Brand groups and economy attributes

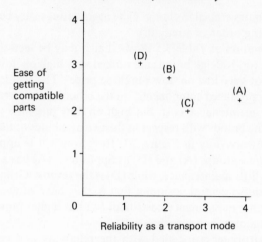

Note: The numbers on the axis are mean consumer ratings.

FIGURE 3.2   Brand groups and reliability/convenience attributes

(D) is perceived to be unreliable as a transport mode but compatible parts for replacements are readily available. Group (B) brands (manufactured by Far East) again occupy a central position with respect to both attributes, that is, bicycles in this group are perceived to be fairly reliable and parts fairly readily available.

When the ratings for all the attributes in each factor are aggregated, the positioning of the four brand groups with respect to the two crucial attitude dimensions tend to be close together. Figure 3.2 also shows quite clearly that brand groups (A) and (B) are very competitive to each other. While group (A) is perceived to be more reliable, group (B) is perceived to be more economical. Moreover, the perceived difference between the two groups in these two crucial factors is fairly small.

As has been noted earlier, there is a marked income effect on the economy dimension while no such effect was observed on the reliability/convenience factor. Thus, given the perceived positioning of the various brand groups on the economy dimension, one should observe a marked difference between consumer income and brand of bicycle purchased. This is further analysed in the next section.

Consumer income and purchase pattern

Table 3.6 analyses the relationship between consumer expenditure and

Table 3.6 Expenditure and brand of bicycle purchased

Household expenditure ($/month)	% consumer who purchased following brand group as their most recent bicycle				Total households
	(A)	(B)	(C)	(D)	
Less than $300	29.4	44.2	8.8	17.6	93
$300 to $499	39.0	27.2	16.9	16.9	72
$500 to $699	38.2	26.5	17.6	17.6	19
$700 or more	46.9	34.3	9.4	9.4	11
Total	34.7	35.7	12.7	16.9	195

the brand group of the most recently purchased bicycle. From the table it can be seen that among low-income families (i.e. with expenditure of less than $300 per month) about 44.2 per cent of them purchased brand group (B), brands manufactured by Far East, with another 29.4 per cent purchasing Raleigh brands. However, for the medium- and high-income families, Raleigh appears to be the favourite brand group. About 39 per cent of the medium-income group (i.e. families with expenditures between $300 and $700 per month) and about 46.9 per cent of the high-income families purchased Raleigh bicycles. There does not appear to be any significant difference in the proportions of households who purchased brand groups (C) and (D) as household income increases. On an over-all basis, brand groups (A) and (B) each account for about 35 per cent of the most recent bicycle purchases by consumers. The dominance of brand groups (A) and (B) over (C) and (D) in market shares is not surprising since these two latter groups have been perceived to be inferior to groups (A) and (B) in respect of the two crucial dimensions affecting choice.

The switch of the dominant brand group from Raleigh among the high- and medium-income households to Far East brands among the low-income families is significant. This confirms the findings of section 3.3 that indicated that, among low-income households, economy tends to be the main consideration in the choice of a bicycle. Among the medium- and high-income families, the relative importance of economy *vis-à-vis* reliability/convenience is less dominant and this is reflected in the greater proportion of these households purchasing Raleigh brands rather than Far East ones.

In the next section, we analyse the technology used by Raleigh and Far East in the manufacture of their bicycles. The income distribution among consumers of these brand groups is then related to the manufacturing techniques applied in order to arrive at some policy implications for promoting the development of manufacturing techniques that meet the basic needs of the majority of the consumers.

3.4 Appropriate Technology for Bicycle Manufacturing

The bicycles manufactured by both Raleigh and Far East include some self-manufactured components and some purchased from external suppliers – locally as well as abroad. Table 3.7 presents details of the origins of the components used in the Raleigh and Far East bicycles. The table also lists the components manufactured by six local bicycle component firms. These firms are fairly representative of the bicycle ancillary industry which consists of about eleven ancillary firms altogether and should therefore be sufficient for the purpose of generalisation regarding the ancillary sector.

In this section, we focus our discussion on a comparison of the manufacturing techniques of Raleigh (a foreign-based firm) and Far East, as well as those used by the other local bicycle component manufacturers. The essence of the comparison is modern manufacturing techniques (i.e. those used by Raleigh) and less-modern techniques (i.e. those used by Far East and other local establishments).

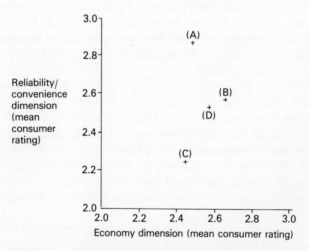

Figure 3.3   Brand groups and the two crucial dimensions

TABLE 3.7   Bicycle components manufactured by primary firms and ancillary components firms

Bicycle components	Primary firms		Local bicycle component manufacturer*					
	Raleigh	Far East	Tan Lian	Sin Heng	Joo & Co.	Gnee Lee	Batu Pahat	Arm-strong
1 Body frame	SM	SM					x	
2 Front fork	SM	SM	x				x	
3 Mudguard	P-UK	P-Lo	x					
4 Crank	P-UK	P-J&T						
5 Chain wheel	P-UK	P-J&T						
6 Chain	P-Lo	P-Lo						
7 Pedal	P-UK	P-J&T						
8 Handlebar	SM	SM					x	
9 Handlebar grip	P-UK	P-J&T						
10 Brake system	P-UK	P-J&T						
11 Hub (front and back)	P-UK	P-J&T						
12 Spokes and nipples	P-Lo	P-Lo		x				x
13 Rims	P-Lo	SM						
14 Saddle	P-Lo	P-J&T	x					
15 Chain guard	SM	P-Lo	x			x		
16 Luggage carrier	P-UK	P-Lo			x	x	x	
17 Bicycle stand	P-UK	P-J&T			x	x		
18 Tyre and tube	P-Lo	P-Lo						
19 Rugs (joints)	P-UK	SM						
20 Rear reflector	P-UK	P-J&T						
21 Bell	P-UK	P-J&T						
22 Locks	P-UK	P-J&T			x	x	x	
23 Lamp	P							
24 Pump	P-UK	P-J&T						

* Only manufacture those items marked 'x'.

Note: SM – self-manufactured.
P-UK – Purchased from UK.
P-Lo – Purchased locally.
P-J&T – Purchased from Japan & Taiwan.

Comparison of techniques

The comparison from the technical point of view of the manufacturing techniques used by Raleigh, Far East and the relevant ancillary firms is summarised in Table 3.8. This shows the main differences between the Raleigh and non-Raleigh manufacturing techniques in terms of input

TABLE 3.8 Manufacturing operations used by Raleigh (Malaysia) and Far East and ancillary firms

Item	Raleigh manufacturing operations	Far East manufacturing operations	Typical ancillary firm manufacturing operations
Body frame and front fork	1 Cut tube to size (automatic) 2 File tube ends (automatic) 3 Cut threads on ends (automatic) 4 Join tubes using lugs (manual) 5 Weld and braze joints (semi-automatic) 6 Electrostatised paint applied by dipping, then components are stove enamelled	1 Cut tube to size (semi-automatic) 2 File ends and cut thread (semi-automatic) 3 Join tubes using lugs (manual) 4 Weld joints (manual) 5 Paint components using ordinary paint and spray system	Same as Far East except paint is applied manually instead of the spray system
Handlebar	1 Cut tube to size (automatic) 2 Bend tube appropriately (automatic) 3 Cut threads (automatic)	1 Cut tube to size (semi-automatic) 2 Bend tube (semi-automatic) 3 Cut threads (semi-automatic)	Same as Far East

4 Join tubes (manual)
5 Weld and braze joints (semi-automatic)
6 Chrome (to acceptable Raleigh standard) handlebar (automatic)

4 Join tubes (manual)
5 Weld joints (manual)
6 Chrome handlebar (automatic)

Chain guard
1 Cut sheet to appropriate size (automatic)
2 Press to appropriate shape (automatic)
3 Join all parts (manual)
4 Weld and braze joints (semi-automatic)
5 Electrostatised paint applied by dipping; components are then stove-enamelled

1 Cut sheet to appropriate size (semi-automatic)
2 Press to right shape (automatic)
3 Join all parts (manual)
4 Weld joints
5 Paint component manually

Rims
1 Cut sheet to length (automatic)
2 Bend and form sheet to circular shape (automatic)
3 Weld ends (automatic)

materials, the basic machining operations of cutting, filing, bending, pressing and threading, and the joining and finishing operations.

Input materials: all input materials used by Raleigh (Malaysia) are imported from its parent company in the UK. For the non-Raleigh firms, most input materials are purchased locally (e.g. steel tubes, bolts and nuts, etc.). These materials may be of acceptable quality but are unlikely to be as good as the Raleigh materials. Specialised items like high-carbon steel sheets are imported from Japan or Taiwan.

Machining operations: in the case of Raleigh (Malaysia), the basic machining operations of cutting, filing, bending, pressing and threading are performed automatically by equipment, resulting in a consistently high quality of product. These operations are performed only semi-automatically by the non-Raleigh firms.

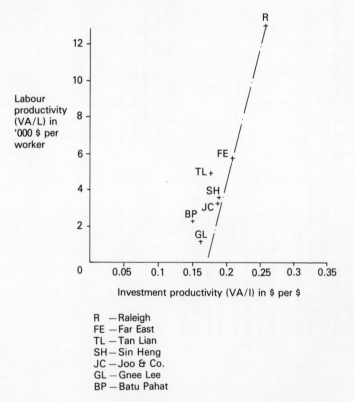

FIGURE 3.4   Relative efficiency of bicycle manufacturing techniques of different firms

It is also evident from Table 3.8 that the techniques of some firms are more efficient (with respect to labour and investment productivity) than those of others. Figure 3.4 shows labour productivity and investment productivity for different firms. The figure shows that the technique used by Raleigh is the most efficient. The technique used by Far East can be considered to be the second most efficient since it dominates all the other techniques except that used by Raleigh. The techniques used by the other firms are inferior to those of both Raleigh and Far East.

In terms of purchases of input materials and components, Table 3.9 shows that, although there is no significant difference between Raleigh and non-Raleigh firms in terms of value of local components purchased per dollar of value added, there is a significant difference between the two groups of firms with respect to value of imported component per dollar of value added. Raleigh uses about $2.1 of imported component per dollar of internal value added, but the average for the non-Raleigh firms is only about $0.60. In terms of the total cost of purchased parts and components, Raleigh uses about $3 worth of externally manu-factured components for every $1 worth of internal value added, while for non-Raleigh firms, the ratio is about $1 externally manufactured components to $1 internal value added. The high value of imported components (and hence of total externally purchased components) may be owing to the fact that Raleigh-imported components are more expensive (and presumably of higher quality) than the imported components utilised by non-Raleigh firms.

Basic needs and the bicycle

The bicycle may be regarded as a product satisfying a basic need in accordance with the Programme of Action adopted by the 1976 World Employment Conference. The basic need that the bicycle satisfies is mobility – the ability to travel from the place of residence to the place of work, and the ability to move about for social purposes.

As was shown in section 3.3, the two crucial characteristics affecting a consumer's choice of a bicycle are economy and reliability/convenience. Figure 3.5 represents the positions of the various bicycle brand groups in the two-dimensional space of economy–reliability. From the figure it is clear that groups (C) and (D) are each dominated by either (A) or (B), implying that rational consumers regard brand groups (C) and (D) as satisfying the basic need for economic and reliable transportation inefficiently. Taking about 20 per cent purchase to be the cut-off point, Table 3.9 indicates that the purchase pattern can be explained by having

TABLE 3.9 Economic indicators of Raleigh and non-Raleigh manufacturing firms

Indicator*	Raleigh	Non-Raleigh firms					
		Far East	Tan Lian	Sin Heng	Joo & Co.	Gnee Lee	Batu Pahat
1 Current value of investment ($ million) (I)†	8.9	5.6	2.2	0.2	0.4	0.3	0.8
2 Employment (L)	194	210	81	20	23	39	49
3 I/L ($ per person)	46 000	27 000	27 000	10 000	17 000	8 000	17 000
4 Value added (VA)/L ($ per person)‡	13 000	5 500	5 000	2 000	3 500	1 300	2 700
5 VA/I	0.29	0.20	0.18	0.20	0.20	0.16	0.15
6 Value of imported components/VA	2.2	0.6	0.6	0.5	0.4	0.4	0.6
7 Value of local purchased components/VA	0.8	0.6	0.4	0.5	0.6	0.6	0.4
8 Capacity utilisation rate (%)§	56.0	56.0	56.0	50.0	50.0	56.0	50.0
9 Advertising expense/VA	0.04	0.008	0.005	0.001	0.001	0.002	0.002
10 Average annual production (in '000 bicycles)	50.0	50.0	—	—	—	—	—

* The relevant base period is 31 December 1978.
† Computed from an appreciation rate of 10% per annum based on cost at time of purchase. Time of purchase of machinery and cost were extracted from the relevant company annual reports. This item also includes working capital which is estimated to be the difference between current asset and current liability.
‡ Labour productivity is computed in $ per worker rather than bicycles per worker. This is because the firms manufacture only parts of a bicycle, and it is not possible to compare their output on the basis of units of bicycles.
§ Based upon a 100 per cent capacity of sixteen working hours per day for 290 working days per year.

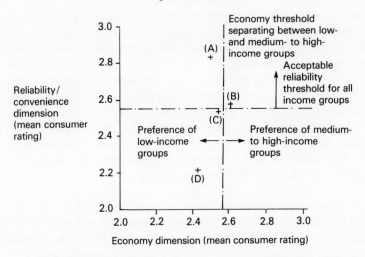

FIGURE 3.5   Positions of brand groups and consumer preference

a reliability threshold line separating brand groups (C) and (D) (i.e. groups (C) and (D) are below, while groups (A) and (B) are above the threshold of acceptable reliability). There also appears to be an economy threshold band which separates groups (B) and (A) with the lower-income families (below $300 expenditure per month) preferring brand group (B) and medium- and high-income families preferring brand group (A).

What Figure 3.5 shows is that, with respect to the brands of bicycles available in Malaysia, brand group (B) (Far East brands) seems to be the 'appropriate' product for families with expenditure of less than $300 per month, while brand group (A) (Raleigh brands) seems to be the 'appropriate' product for families with expenditure of more than $300 per month. Raleigh brands embody the degree of reliability and quality (at the expense of economy) preferred by medium- and high-income families. Far East brands sacrifice some reliability and quality (though the degree of reliability is still above the threshold level) for economy, which is preferred by low-income families.

Appropriate technology

Setting aside the concept of consumers' sovereignty for the moment, and recalling Morawetz's definition of appropriate technology,[12] it is clear that we must view the appropriateness or inappropriateness of a technology from a perspective broader than consumers' sovereignty. In this context, Stewart has suggested several desirable characteristics for an appropriate technology.[13] These include not-too-high an investment/labour ratio requirement, a medium scale of operation, relative independence from foreign inputs in terms of materials and expertise, simplicity of operation and maintenance, etc. From this perspective, it is clear that the technology utilised by Far East (with improvements in peripheral operations and control, to ensure a more consistent output) can be considered as the 'appropriate' technology for bicycle manufacturing in Malaysia. The reasons for this observation are:

1. Since it is an improved version of the present Far East technique, it could be relatively efficient in the sense that no other technique currently available in the country can achieve the same level or labour productivity with the same or lower level of investment productivity.

2. Among the bicycle-manufacturing techniques, its value of investment intensity is intermediate between the highly investment-intensive Raleigh technique and the highly labour-intensive (but inefficient) techniques of the bicycle ancillary firms. In relation to the industrial sector as a whole, this technique's requirement for investment is 'reasonable' in the sense that it is far below the investment intensity of sectors that are obviously investment-intensive (e.g. petrochemicals) and above sectors that are obviously labour-intensive (e.g. footwear manufacture). Thus, its demand on the country's capital-investment resources (limited as they may be) seems to be consistent with its employment-generating potential.

3. In terms of scale of operation requirement, that required by Far East is not beyond the reach of the currently smaller firms or that of the larger ancillary firms. In terms of organisational requirement, it is certainly not beyond the ability of the entrepreneurs who run the ancillary firms, since the basic style of organisation is the same. In terms of the skill requirement of the workforce, the types of skilled workers required are not very different from those in the ancillary firms. All in all, this technique seems to be easy for local entrepreneurs to adopt.

4. The Far East technology manufactures a product that can meet the basic need for reliable transportation at a price acceptable to low-income families. For these families, the bicycle is the only feasible form of family-owned transportation. The technique of production is therefore consistent with the nation's developmental priority which is the eradication of poverty.

5. Thus, there is a considerable potential for bicycles manufactured by the Far East technique. The characteristics of the bicycle seem to represent a good compromise between low purchase price and low maintenance cost, two important attributes of the economy factor, and reliability and convenience, two important attributes of the reliability/convenience factor. With a more co-ordinated marketing and advertising campaign emphasising the reliability and convenience aspects of the product, it can probably make bigger inroads into the medium- and high-income markets as well as increase its share in the low-income market.

Remarks on selected appropriate technology

It should be noted that the suggested appropriate technology (appropriate with respect to the national developmental goal), which is an improved version of Far East technique, is one based on new equipment. The characteristic that makes it appropriate is essentially the substitution of labour in the machine-peripheral activities. For example, in the cutting/filing/threading operation for body frame manufacture, the tube length is manually adjusted instead of being done by machine. The basic 'core' machine activities on which the basic reliability of the product depends are basically similar to those of Raleigh (the most investment-intensive technique).

3.5 The Pattern of Poverty, Appropriate Technology and Employment Generation

This section contains a brief discussion of the pattern of poverty in Malaysia. We then examine how the utilisation of an appropriate technology for bicycle manufacturing is consistent with the government's developmental goal of the eradication of poverty by improving the standard of living of the lowest income group in the population.

Malaysian poverty pattern

In terms of the absolute number of poverty households,[14] Table 3.10 shows that the total number of poverty households increased from 791 800 in 1970 to 835 100 in 1975. In 1975, it was estimated that poverty occurred in 54.1 per cent of the rural households but in only 19 per cent

TABLE 3.10 Peninsular Malaysia: poor households by rural and urban strata, 1970–90

Strata	1970		1975		1980		1990	
	Total poor households ('000)	% of total households	Total poor households ('000)	% of total households	Total poor households ('000)	% of total households	Total poor households ('000)	% of total households
Rural	705.9	58.7	729.9	54.1	646.7	43.1	388.9	23.0
Urban	85.9	21.3	105.2	19.0	121.6	15.8	125.0	9.1
Total	791.8	49.3	835.1	43.9	768.3	33.8	513.9	16.7

Source. Malaysia, *Third Malaysia Plan, 1976–80* (Kuala Lumpur: Government Printer, 1976) p. 73.

of the urban households. The incidence of poverty among rural households is particularly high, and government developmental strategies aimed at reducing the number of poor households to 43.1 per cent of the rural households and 15.8 per cent of the urban households in 1980, and to 23 per cent and 9.1 per cent of the rural and urban households respectively in 1990. In terms of absolute numbers, these plans called for the reduction of poor households to 768 300 in 1980 and 513 900 in 1990.[15]

Given the importance attached by the government to raising the level of income of the poorest income group above the poverty line, there will be a higher and higher proportion of the population above the poverty line but still below the medium- and high-income line by the end of the various development plans.

In 1967–8 (Table 3.11), 55.3 per cent of all households earned income of less than $100 per month, which was the generally accepted poverty line at that time.[16] About 35.7 per cent of the households earned $100–$299 per month, corresponding to the households above the poverty line but below the medium- and high-income levels. From Table 3.11 it can

TABLE 3.11 Ownership of consumer durables by household income

Type of household item	% of households with following household income ($/month) owning:*			
	1–99	100–299	300–749	$750 and over
Motor-car	0.3	5.2	45.7	78.9
Motor-cycle/scooter	2.8	17.8	24.8	10.2
Bicycle	47.1	72.3	59.7	37.8
Telephone	0.1	1.8	14.5	50.1
Radio	23.1	53.4	80.8	84.5
Television	0.3	7.0	39.5	43.3
Refrigerator	0.3	5.2	46.3	83.1
Electric fan	0.8	13.6	57.7	86.0
Air-conditioner	0.1	0.2	1.5	19.1
Sewing machine	22.4	57.9	79.6	69.1
Total households in income group ('000)	899.2	580.6	115.1	30.2
% of total households	55.3	35.7	7.1	1.9

* Income here is defined to be cash income.

Source. Malaysia, *Socio-Economic Sample Survey of Households, Malaysia, 1967–68: Household Amenities and Convenience, West Malaysia* (Kuala Lumpur: Statistics Department, 1974).

be seen that the bicycle is the major item of consumer durables owned by these two groups of households. What Table 3.10 emphasises is that, for families in poverty and just above poverty categories, the bicycle is a very important consumer durable. It is owned by about half the families in the poverty category and about three-quarters of the families in the just-above-poverty category. Hence, if the government targets – as set out in Table 3.10 – for the eradication of poverty are achieved, there will be a tremendous increase in demand for bicycles of the type manufactured by the Far East-type technique – bicycles that are cheap but reliable enough to meet the basic transportation needs. Given this expected large increase in demand for such bicycles (i.e. bicycles manufactured by the appropriate technique), it is not only socially desirable but also consistent with the objective of the raising of the living conditions of the poor, that government policies on the bicycle sector should be directed towards the growth of manufacturing establishments utilising the appropriate technology, that is, Far East-type technology that is efficient but not too highly investment-intensive.

Appropriate technology and employment generation

The encouragement of the utilisation of appropriate technology for bicycle manufacturing will also have the desirable effect of generating more employment opportunities than if the more investment-intensive Raleigh-type technology were to be encouraged. The current average annual production of both Raleigh and Far East is about 50 000 bicycles each and there may be scope for two more plants of the present Far East capacity by about 1981.[17] The required investment for these two plants, if the Far East-type technique is used, would be about $11.2 million with an employment potential of over 420 workers. However, if the Raleigh-type technique is adopted for these two plants, the required investment is about $16.8 million with an employment capacity of about 390 workers. Hence, with the Far East-type technique, there would be an extra thirty jobs created. Moreover, the surplus investment of $5.6 million (i.e. $16.8–$11.2 million) can be invested in other sectors to create more employment.

Government policies and appropriate technology

The current Malaysian scheme for the promotion of industrialisation (as incorporated in the Investment Incentive Act of 1968 and the Investment Incentives (Amendment) Act of 1971) is aimed at promoting the growth of firms using technology that is either very investment-intensive or very labour-intensive. Appropriate technologies generally have an intermediate degree of factor intensity. Some fine tuning of the incentives is necessary if appropriate technology is to be promoted. There need to be incentives based on both fixed investment and employment.

Implications for policies to promote the application of appropriate technology

This study highlights the role of the tariff structure in the promotion of technologies, which utilise more intensively the factor endowment the country is blessed with. While not arguing totally against tariffs, the study finds that the tariff structure ought to be designed so that products or components that utilise more labour-intensive processes are protected against international competition. Appropriate technologies should be adopted for the manufacture of products not only for the domestic market but also for the international market where the real test of their 'appropriateness' lies.

Currently, the Malaysian industrial incentives are such that the most investment-intensive (or the most labour-intensive) technology receives the most encouragement. This tends to discriminate against appropriate technologies, which are generally intermediate in terms of both investment and labour intensity. The Malaysian Industrial Development Authority should consider the possibility of granting fiscal incentives based on both employment and fixed investment. Such a schedule is not difficult to construct. If it is offered together with the other incentives, it would go a long way towards the promotion of appropriate technologies.

Besides restructuring the incentive scheme, the government should consider the possibility of conducting more research, or encouraging it to be done, on the development and identification of appropriate technology on a product-by-product or sector-by-sector basis. This study demonstrates the usefulness of such investigations. The research need not be conducted for all products or all sectors of the economy but only for the products or sectors deemed relevant from the viewpoint of national development. Such research may provide insights into technology adoption and assimilation of local entrepreneurs, and how these processes can be modified or harnessed to enhance the development of appropriate technologies.

The government should offer consultancy services and advice to local entrepreneurs not only on problems such as finance and operations, but also on aspects of marketing and product-image development. This study demonstrates that foreign-based firms, sophisticated in their marketing and advertising strategies, can create non-basic needs among consumers. It should be the function of government to advise other firms on advertising strategies.

At a macro level, there should be a national technology policy on the role and promotion of appropriate technology and the relationship of technology in general, and appropriate technology in particular, with the nation's development goals. The policy should focus on issues like the rapid and efficient transfer of foreign technologies, its adaptation to the local environment and channels for the dissemination of these adapted technologies. While not calling for the establishment of a full-fledged technology development institute, the government should nevertheless strive for a high level of scientific understanding among its high school and university students. This would encourage the quick adoption and assimilation of foreign innovations and promote the rapid evolution of appropriate technologies. The current heavy emphasis on science education in the high school system in Malaysia is certainly a major step in the right direction.

Notes

1. Associate Professor, Faculty of Economics and Administration, University of Malaya, Kuala Lumpur.
2. Because of the ease with which the bicycle could be assembled, the bicycle was usually imported in the form of components and assembled in the bicycle shops.
3. Effective from 5 October 1978. The monetary unit used in this report is the Malaysian dollar (roughly US$0.45). An average locally manufactured bicycle is retailed in the range of $130–$220 per unit. Hence the tariff of $60 per imported unit represents protection of between 45 per cent to 25 per cent of the retail price – high by any standard.
4. This figure was derived from D. Lim, *Economic Growth and Development in West Malaysia 1947–1970* (Kuala Lumpur: Oxford University Press, 1973) p. 285.
5. There is at least one bicycle dealer in every town with a population of over 10 000.
6. In fact, very little (if any) of the bicycle trade in Malaysia is done through the informal marketing system, that is, through retailers and outlets that are not registered or identified by the government.
7. See, for example, P. E. Green and D. S. Tull, *Research for Marketing Decisions* (Englewood Cliffs, N. J.: Prentice-Hall, 1975).

8. Malaysia, *Census of Population and Housing 1970*, (Kuala Lumpur: Statistics Department, 1974).

9. These attributes were arrived at after thorough research on consumer attributes of bicycle and similar products (e.g. motor-cycle and motor-car) and are considered to be fairly comprehensive. See also P. Green, S. Maheshwani and V. Rao, 'Dimensional Interpretation and Configuration Invariance in Multi-Dimensional Scaling: An Empirical Study', *Multivariate Behavioural Research*, vol. 4 (1969) pp. 159–80.

10. This was computed on the basis of the following weights: crucially important (4), highly important (3), important (2), and somewhat important (1). Hence the lower the mean score, the more important is the attribute.

11. For a self-employed household, its income would include an estimation of its own home-grown food consumed, changes in its inventory holdings, costs of production, etc. These implications make estimation of household income extremely difficult. See also W. van Ginneken, *Rural and Urban Income Inequalities* (Geneva: ILO, 1976).

12. See D. Morawetz, 'Employment Implications of Industrialisation in Developing Countries: A Survey', *Economic Journal*, vol. 84 (335) (Sept 1974).

13. See F. Stewart, *Technology and Underdevelopment*, (London: Macmillan, 1977).

14. The poverty line income is the income necessary to cover minimum nutritional requirements and essential non-food expenses so as to sustain a decent standard of living (Malaysia, *Third Malaysia Plan 1976–80* (Kuala Lumpur: Government Printer, 1976).

15. It is not the intention of this study to evaluate the feasibility (or otherwise) of the achievement of these targets. Hence these targets are accepted at their face values.

16. The *1967–68 Socio-Economic Sample Survey* defined income to be cash income. This generally understates income of the low-income groups, since these groups generally supplement their cash income with home-produced foodcrops, transfer payment from relatives, etc.

17. The projected demand by 1981 is about 196 000 bicycles per year. Hence there will be enough scope for four bicycle plants of capacity of 50 000 units per year. Again we are making the conservative assumption that Far East-type technology requires at least a workforce of 200.

4 Metal Household Utensils and Basic Needs in India

T. S. Papola and R. C. Sinha[1]

This study attempts to examine the interrelationships between the pattern of income distribution, consumption, production technology and employment, in the case of an individual commodity group, namely metal utensils. Metal utensils do not seem to have attracted as much attention as an item of basic needs as food, clothing and shelter. No doubt metal utensils form a small item in terms of total household consumption expenditure; and in certain regions and low levels of incomes, lower-order products such as utensils made of earthen clay are also used to serve the needs of the household. But even in these situations certain functions require the use of metal utensils and earthen ones fail to fulfil certain minimum requirements of durability, ease of handling and cleanliness. Therefore, metal utensils essentially belong to the category of goods that not only serve certain basic needs of a household, but also have the characteristics of minimum essential items of household consumption.

It is difficult to lay down norms for basic-needs goods such as furniture and household equipment, while it is relatively easy in the case of food, clothing and shelter. The dimension that needs to be taken into account here relates to the quality and characteristics of the products. Commodities that constitute basic needs are consumed both by poor and non-poor; but their characteristics tend to differ. The items consumed by relatively higher-income groups are believed to have characteristics that are not essential for the fulfilment of basic needs as they may be more appealing in terms of sophistication, design, finish, etc. They are, therefore, produced at higher cost, have higher prices, and generally use capital-intensive methods of production. Moreover, these items may be the same as identified to constitute basic needs, but the structure of their characteristics may be more suitable for the non-poor which would raise their price to a level that the poor cannot afford; they may also be produced with the help of technology that uses less labour thus reducing the income-earning opportunities for the poor.

Another problem with analysing metal utensils is the behaviour of change in expenditure of this item in relation to household income and expenditure levels. Being a consumer durable, the purchase pattern of metal utensils in the short period is, of course, bound to be different from that of current consumption items like food. But while households at low-income levels may spend more on this item with a rise in incomes for meeting the shortfalls in their current requirements, those with higher incomes may also spend proportionately higher amounts for the sake of variety and on items of only casual use. Thus the income elasticity of demand for metal utensils, despite its being an item of basic need, may continue to be high even beyond the fulfilment of basic needs.

Metal utensils, even though a small item of household consumption in relation to the total expenditure, thus provide an interesting case for examining the various facets of the income distribution–consumption–production–technology–employment relationship. The specific aspects that we shall examine in this chapter include: (i) an examination of the use pattern, composition and current demand of metal utensils used by the households in different income groups, (ii) identification of the relative and absolute importance of the variety of metal utensils in the household budget, (iii) an examination of the extent of and constraints on the fulfilment of basic needs under existing structures of demand, production technology and supply conditions, (iv) an overview of the effect of changes in the levels and distribution of income on employment in the metal utensils sector, and (v) indication of the options to regulate production technology and supply conditions of metal utensils with a view to augmenting employment potential and fulfilment of basic needs.

In view of the non-availability of any systematic data about consumption, production and marketing of metal utensils, the study uses data collected through primary investigations about rural and urban household consumption patterns, production technology and marketing of metal utensils. The field study for collection of requisite data was carried out in selected rural and urban areas in the State of Uttar Pradesh (India). The rural consumption survey was carried out in five villages (Godhana, Mirzapur, Ahmadpur, Ataria and Manwan) of Sitapur and four villages (Deorai, Kalau, Saramau and Mampura) of Lucknow district. Some of these villages are within the catchment of the Lucknow–Sitapur highway while others are remotely situated. The selection of sample households was done on the basis of probability proportionate to size (PPS), from among agricultural labourers, various categories of farmers identified by landholding size and rural artisans. The urban sample was drawn randomly from the various localities of Lucknow city, by identifying the households by the type and location of their residences. The total size of the sample was 600 households – 400 in rural and 200 in urban areas.

For collecting information on technology and production conditions, interviews were held with twenty-three metal-utensil manufacturing units in Kanpur, Lucknow, Moradabad, Basti and Ghaziabad, engaged in the manufacturing of iron, aluminium, brass, *phool* and stainless-steel utensils respectively. Information about marketing of utensils was collected from wholesalers and retailers in the city of Lucknow and also from the retailers in the regular and weekly markets of the rural areas along the Lucknow–Sitapur road. The total number of trading units surveyed is twenty-five, including six wholesalers. The reference year for data collected was 1979–80.

While every precaution was taken to ensure correctness and reliability of data collected from households and production and sales units, certain points need to be noted in respect of problems faced during the survey. First, we suspect a slight underestimation of the stock of metal utensils with the households in the rural areas on account of the respondents' hesitation in disclosing the quantity of metal utensils; they are sometimes considered as valuable assets, and the respondents may be afraid that the knowledge about the utensils would prompt thefts. Secondly, one is also not always sure about the money value of the stock of metal utensils, particularly old ones, some of which could have been purchased a long time ago or even inherited. It may be pointed out, however, that for the main analysis we have utilised the information relating to utensils in regular use only which are generally of relatively recent acquisition and, therefore, their recorded prices are not likely to be very much different from the actual prices. Thirdly, a number of production units contacted for the collection of data on production technology were small sized and operating on a household basis, particularly in the case of brass and *phool* utensils. These units did not have systematically maintained records of their production and, therefore, the information supplied was primarily based on their memory and experience. The number of workers employed by these units varied from day to day depending on the volume of work and the number of processes to be handled. In such a situation, the information on average number of workers and man-days is likely to involve some degree of error. Fourthly, the problem of lack of record and therefore reliance on memory was also encountered in the case of small trading units.

4.1 Structure of Consumption

Commodity characteristics

Utensils as a commodity group represent a variable mix of items with differing shapes, sizes, material contents and quality differences; and these characteristics are not necessarily related to different use categories. In principle, it would be best to analyse the basic-needs content of utensils in terms of durability and efficiency. In practice, however, we shall analyse demand in terms of use categories, such as cooking, serving, carrying and storage and quality characteristics will be taken care of by incorporating metal categories in the analysis. In addition, we opted to examine the technological conditions of production on the basis of the metal content of utensils. The metals used are: iron, aluminium, brass, *phool* (an alloy of tin, copper and nickel), *kaskut* (a mix of metal scraps including *phool* scraps), copper, German silver (an alloy of copper, zinc and nickel), and stainless steel. The non-metal materials used for making utensils are china clay, glass, ordinary clay and plastics. For certain kinds of uses like serving and storage, metal and non-metal utensils are substitutable for one another. The present study, however, focuses on metal utensils only. In any case, it is the metal utensils that are used for the most essential of the requirements, namely cooking, and according to our estimates the metal group accounts for around 98 per cent of the value of all kinds of utensils possessed and purchased by the households.

The metal utensils can be classified into the following five use categories: cooking (C), cooking accessories (CA), serving (S), serving accessories (SA) and storage and carrying (SC). The following description gives a qualitative idea of the structure of utensils generally used by households in India for serving different functions.

Utensils that are mostly and almost exclusively used for cooking purposes are *tawa* (flat frying-pan), *karhai* (deep frying-pan), *batuli* (round-shaped boiling pot); and cooking accessories, namely *karchhul* and *chamcha* (both cooking spoons), and *chimta* and *sansi* (both forks used for holding heated utensils). Other cooking utensils are *tasla* (a hollow and flat-based utensil, generally bigger than full saucer), *bhagona* (a boiling pan), kettle, frying-pan, cooker, pressure cooker, etc. Within the cooking category it is also possible to distinguish between utensils that are directly and indirectly used in the cooking process. An indirectly used cooking utensil is, for instance, *tasla*, in which flour is processed to make dough for preparing bread.

The most common serving utensils are *thali* (a metal plate), tumbler and bowl. Other serving utensils are tray, cup, mug, spoon, fork, knife, etc., which are relatively less popular among low-income groups, particularly in rural areas. The storage and carrying utensils are *kalash* (a large round-shaped vessel used for storing water), buckets, jars, milk pots, tiffin boxes, etc., and *ghara* and *surahi* (both earthen containers commonly used for storage and carrying of water particularly in rural areas).

The analysis in this section basically aims at the examination of consumption behaviour of the households in respect of metal utensils, in relation to their income levels proxied by per capita expenditure (PCE) levels. Consumption is specified here in terms of stock and current purchases. It may be mentioned that the total demand or requirement of metal utensils as a group item is not reflected fully in the current household purchases. Metal utensils are durable items and are purchased in the event either of the need for replacements or of additional requirements. Thus, the stock of metal utensils is of greater relevance for analysing the consumption of this item, particularly when one is interested in the fulfilment of the basic needs. The impact of income redistribution has, however, been visualised in terms of changes in the current demand (purchase) pattern, assuming that average behaviour of the consumers in a particular income class is uniquely determined.

The hypothesis underlying our analysis in this section implies, on the one hand, that the needs are fulfilled by metal utensils, can be satisfied in different ways representing various degrees of essentiality and non-essentiality, and on the other hand, the various metal bases of utensils represent a hierarchical order based on their costs and prices which also broadly corresponds to the degree of essentiality and non-essentiality of need fulfilment. For example, plates are essential serving utensils, but table spoons may not be considered essential in a situation where people are normally eating with their fingers. An aluminium or brass plate is essential, although costlier, more elegant and durable; stainless-steel plates could be used if one can afford them. Similarly, cooking vessels for staple food and lentils are necessary; a pressure cooker may not necessarily represent the basic-need characteristics so long as housewives have plenty of time to cook with the help of traditional utensils. Finally, the cooking vessels could be of iron, aluminium and steel, though the cheapest among them will be good enough for basic-needs purposes.

It is further assumed that at low-income levels the expenditure on metal utensils would primarily represent the expansion of the stock of utensils with essential characteristics both in terms of use variety and metal category. However, increase in expenditure at high-income levels would primarily represent the variety in need fulfilment and quality, elegance and durability of the utensils stock. It should further be mentioned that the metal utensils can also be purchased, by those who can afford them, for only occasional use. It is not uncommon to find a few spare dinner sets in the possession of middle- and higher-income groups in the urban areas, or of the relatively rich in the rural areas, for use on special occasions by their families and guests. In the villages, social and religious ceremonies require large-size cooking vessels and a large number of serving utensils that are not needed for normal household consumption. For the purposes of fulfilment of basic needs of households, however, it is the stock in regular use that is relevant.

Another important aspect of the analysis of consumption of metal utensils in relation to household income levels, is the quality difference among various items that fulfil the same need. It is, however, difficult to define precisely the qualitative characteristics of utensils except in terms of certain proxies. The branded products are often considered qualitatively superior to unbranded ones, and within the group of branded products it may be possible to rank the products by the reputation of the manufacturing concern. A large proportion of the metal utensils in India are, however, unbranded and produced in the medium, small and very small units in the semi-organised and unorganised sectors. The only variable through which quality variations can be inferred are the relative prices, which could be used with the assumption that a difference in price reflects the differences in metal base weight, production techniques and other attributes of quality.

Household stock of utensils

Leaving out the non-metal utensils that form but a small proportion of total consumption of utensils, we now turn to the consumption behaviour of households in respect of metal utensils. No doubt the expenditure on stock of this item is generally positively associated with the per capita household expenditure level, but the relationship is not always consistent and proportional. Households with a per capita expenditure (PCE) level of less than Rs.350 have a stock of Rs.144 worth of metal utensils in the rural areas, but even in the highest PCE range of over Rs.2200 the figure is only Rs.555. Thus it looks as if in the rural areas the stock of metal utensils rises less than proportionately with the rise in income levels. In the PCE ranges up to Rs.950, the absolute stock of metal utensils has only a small variation between Rs.144 for the lowest and R.200 for the highest expenditure range. It is only in the PCE ranges above Rs.950 that one finds a steeper rise in stocks of metal utensils than at lower-income levels. The inter-group variations are, however, much wider in urban areas; members of the lowest income group have utensil stock worth only Rs.34 while those with Rs.1600 or more per capita income have a stock worth Rs.1200.

Two inferences can be drawn from the details given above: first, the stock of metal utensils is not a monotonous function of income levels; it has discontinuities, particularly after the households reach sufficiently high income levels. In the lower ranges the perception of needs is more or less fixed to the fulfilment of essential requirements. Secondly, the degree of deprivation from fulfilment of basic needs in terms of stock of utensils is probably lower in the rural than in the urban areas, particularly at the lowest ranges of per capita income; but the degree of non-essentiality creeps in more steeply in the high-income ranges in urban areas than in rural areas.

One also finds in the highest rural-income class that about two-fifths of the metal utensils are not regularly used. In other income classes this proportion is on average about one-tenth. Non-regularly used metal utensils are normally inherited or received as wedding gifts.

Use categories.   In the analysis of consumption that follows we shall only consider the regularly used metal utensils. Table 4.1 shows the number and value of regularly used metal utensils possessed by rural and urban households.

The variations in the quantity and value of regularly used metal utensils among households in different PCE ranges in *rural areas* are mainly found among the serving utensils. The consumption of cooking utensils, including cooking accessories, seems more or less fixed among different expenditure groups. It seems that a household requires a minimum of cooking utensils, irrespective of its economic status; additions beyond this minimum are redundant even if a household could afford them. The better-off households can, of course, go in for qualitatively better and costlier utensils, as is reflected in higher value figures in those classes. The range of qualitative variations in rural areas seems, however, rather limited. The number and value of utensils used for carrying and storage of food and water show a consistent, though less than proportionate, increase with increasing expenditure levels. The qualitative change in the consumption of such utensils is virtually absent as the value increases proportionately with numbers, except in the case

TABLE 4.1 Possession of stock of regularly used metal utensils.
Rural and urban households (in number (N) and value rupees (V))

Per capita expenditure (Rs./year)	Cooking		Cooking accessories		Serving		Serving accessories		Storage and carrying		Total	
	N	V	N	V	N	V	N	V	N	V	N	V
Rural areas												
1 Below 350	4.2	33.2	2.1	4.5	5.8	52.9	0.7	0.4	2.9	37.2	15.7	128.0
2 350–425	4.5	43.9	2.7	8.5	8.1	48.2	1.0	1.4	3.4	45.7	19.7	147.6
3 425–500	4.0	41.1	2.4	4.8	6.0	56.0	1.5	1.9	2.7	41.0	16.6	144.8
4 500–600	4.3	46.1	2.6	6.6	6.9	46.7	0.4	1.4	3.0	44.9	17.1	145.7
5 600–750	4.4	59.5	3.0	7.5	6.8	58.7	1.3	8.8	3.5	54.6	19.0	189.2
6 750–950	4.7	56.7	3.2	8.5	7.5	77.9	0.7	0.8	3.3	53.4	19.2	197.3
7 950–1250	4.4	58.6	3.5	13.2	7.3	68.5	1.3	5.9	4.1	68.9	20.6	215.1
8 1250–1600	5.5	75.1	3.7	10.0	10.4	107.3	2.5	4.1	4.5	79.6	26.6	276.1
9 1600–2200	6.1	69.0	4.1	10.9	11.6	138.6	2.8	0.7	5.0	77.4	29.6	296.5
10 2200 or over	4.7	56.5	3.0	8.0	11.3	11.4	4.7	4.7	4.3	133.0	28.0	316.2
All classes	4.6	54.2	3.0	8.3	7.6	69.1	1.2	3.3	3.5	56.0	19.9	190.8

92

Table 4.1 cont.

Per capita expenditure (Rs./year)	Cooking		Cooking accessories		Serving		Serving accessories		Storage and carrying		Total	
	N	V	N	V	N	V	N	V	N	V	N	V
Urban areas												
1 Below 350	1.8	9.2	2.8	8.3	5.5	16.5	–	–	–	–	10.1	34.0
2 350–425	–	–	–	–	–	–	–	–	–	–	–	–
3 425–500	2.9	54.4	3.7	22.1	4.4	44.1	3.7	5.9	3.7	42.7	18.4	169.1
4 500–600	4.9	38.0	1.2	3.7	7.4	41.1	1.8	0.9	1.2	19.6	16.6	103.4
5 600–750	6.5	121.0	3.9	20.5	11.4	154.2	3.2	5.8	2.5	42.5	27.3	343.9
6 750–950	6.4	75.7	4.0	20.9	10.4	82.3	4.2	5.0	3.7	49.4	28.6	233.3
7 950–1250	6.3	134.5	3.9	18.5	12.9	197.2	3.2	5.5	3.5	59.9	29.7	415.6
8 1250–1600	8.8	205.9	5.1	29.3	15.2	182.2	6.4	11.8	4.0	63.6	39.5	492.7
9 1600–2200	9.6	466.9	5.1	26.7	30.8	556.7	12.3	27.1	4.7	64.5	62.7	1 142.0
10 2200– or over	5.3	342.6	3.7	23.2	36.3	709.5	9.5	23.7	4.2	84.2	59.0	1 183.2
All classes	6.9	164.6	4.2	21.7	14.6	203.5	5.0	9.0	3.5	55.1	34.1	453.9

of the highest PCE range (Rs.2200 or over) where the value of storage and carrying utensils rises abruptly though their number is only slightly above average. The serving utensils show a significant rise in number beyond the per capita income range of Rs.1250, after rising only gradually with expenditure levels in the lower ranges. Increase in value is generally proportionate to that in number of utensils in this category, implying very little qualitative variations.

In *urban areas* utensils in all use categories show a steep rise in number as well as value with rising PCE ranges. This is so even in the case of cooking utensils; in the lowest expenditure group households have only two cooking utensils with an average value of Rs.9.2 while those in the per capita expenditure range of Rs.1600–2200 have almost ten such utensils with a value of Rs.467 on average. If the rural average of 4.6 cooking utensils per household is taken as the necessary minimum, urban households in the per capita expenditure range up to Rs.500 fall short of this requirement and those in the PCE ranges between Rs.1250 and Rs.2200 have twice the number of utensils than the minimum required. Leaving apart the lowest and some of the highest ranges, the value of cooking utensils in the urban areas rises almost in line with their number among the different PCE ranges, thus implying little qualitative variation. In the case of cooking accessories the number as well as the value does not seem to change much across the expenditure groups.

The variations are very wide both in number and value of serving utensils though not so much in serving accessories. In the PCE ranges up to Rs.600 per capita, the urban households have five to seven items in this category, those in the Rs.600–1250 ranges ten to fifteen items, and those in the higher ranges have thirty serving utensils on average. The value of these utensils changes from the first to the second group in proportion with the number, but much more sharply in the third group compared with the second. Thus those in the middle-income ranges buy larger stocks of similar serving utensils than do the lower-income groups, but the high-income groups buy a larger number of qualitatively superior serving utensils. The carrying and storage utensils do not vary widely among the different income ranges either in number or value; and despite the wide differences in the level of consumption both in terms of number and value of metal utensils, the rural–urban differences in the consumption of this category of utensils are also not significant.

The above description of the household consumption behaviour according to use categories suggests that cooking utensils form the foremost category of utensils for serving the basic needs of the households. The degree of non-essentiality in the household expenditure on this item is therefore the lowest, though the highest income brackets in urban areas do seem to have some stock of cooking utensils which may not be necessary for the fulfilment of basic needs. This is also the case with storage and carrying utensils, where non-essentiality in quantity terms seem non-existent, though the highest income group in rural areas seem to have included a good degree of quality non-essentiality in their stock of these utensils.

It is primarily the stock of serving utensils that includes a large number of items not relevant for the fulfilment of basic needs, both in rural and urban areas. It is not only that the households in the high-expenditure ranges buy more than the necessary number of serving utensils but their purchases are also of relatively better quality and higher value.

Metal categories.   The different metals forming the material base of utensils differ in terms of weight per unit of volume, durability, elegance and price. The five different metals specifically considered in the present analysis are: iron, aluminium, brass, *phool* and stainless steel; the rest of the metals used for manufacturing utensils are clubbed together in the category of 'others'. The residual category of 'others', though claiming a significant percentage of stock as well as current purchases, could not be disaggregated for the present analysis. It consists of a wide variety of metals and alloys (*kaskut*, zinc, German silver, copper and nickel, to name a few) for which information on production conditions was extremely difficult to obtain. Therefore, the analysis has been carried out in terms of one group. There is no direct relation between metal content of a utensil and the use to which it is put; the same metal can be used for the manufacture of utensils of different uses. The metal content of utensils purchased by households from different income classes is therefore a function of the relative prices, along with durability and sophistication. One can generally assume that the qualities of elegance and sophistication will be more important for the higher-income groups, while among the lower-income strata of households the price will prove to be the most important determinant. Durability can outweigh price differentials to some extent even among the lower-income groups.

To a limited extent, however, there is a relationship between on the one hand the use, price, durability and other related attributes of utensils, and, on the other, their material base. While cooking utensils and cooking accessories of iron are used by all classes of people, aluminium is often referred to as the 'poor man's metal'. That does not, however, mean that aluminium utensils are not used by higher-income groups. The use of cooking utensils made of brass is on the decline because brass apparently interacts with food having acidic contents. The use of *phool* utensils is experiencing a steeper decline owing both to supply and demand conditions. In rural areas the possession of *phool* utensils is often still considered as a status symbol. Stainless steel is the new and upcoming metal, but it is also the costliest and not all households can afford it.

In our sample the iron and brass utensils, in that order, are most important in the utensil stock of rural households, each claiming about 28–29 per cent of the total value (see Table 4.2). Aluminium comes next with 19 per cent of the stock value. *Phool* claims 2 per cent and stainless steel 4 per cent, and the rest, about 18 per cent, is shared by other metals. In urban areas, brass is at the top – claiming 34 per cent of the value of stock of household metal utensils. But the next place is held by stainless steel with a 23 per cent share followed by aluminium at 19 per cent and iron at 16 per cent. *Phool* also constitutes a significant, 7 per cent, share of the value of utensil stock of urban households. 'Other' metals are not found to be significant.

Iron, aluminium and brass, holding importance in that order, together make up about 80 per cent of the total value of utensil stock of households with a PCE up to Rs.600 in rural areas. These metals claim over three-fourths of the value of the utensil stock in households with a PCE between Rs.600 and Rs.2200, but the most important metal in this group of households is brass and not iron. In the highest-income group (Rs.2200 or more PCE) stainless steel becomes the most important metal with a 47 per cent share, followed by brass, iron running a poor third. The share of *phool* utensils has no relationship with income ranges, though it does not seem to be popular at the lowest and the highest points of income distribution. Consumption of stainless-steel utensils is insignificant in the rural households with a PCE up to Rs.950; it seems to gain some importance in the PCE range of Rs.950–2200, but becomes the major item in the utensil stock beyond the PCE level of Rs. 2200. The consumption of utensils made of other metals is not related to the PCE levels.

In urban areas, households with a PCE up to Rs.500 mainly use aluminium utensils but in the next PCE range of Rs.500–600, iron, aluminium and brass have an almost equal share in the utensil stock of the households. Households in these ranges did not report the use of stainless-steel utensils. In the PCE ranges between Rs.600 and Rs.1250, brass holds major importance; in the next range, Rs.1250–1600, it shares importance almost equally with iron, aluminium and stainless

TABLE 4.2 Value of current stock of regularly used utensils by metal category. Rural (R) and urban (U) areas (in %)

Per capita expenditure (Rs./year)	Iron		Aluminium		Brass		Phool		Stainless steel		Others		Total	
	R	U	R	U	R	U	R	U	R	U	R	U	R	U
1 Below 350	36.0	10.8	28.9	89.2	15.8	—	—	—	0.4	—	18.9	—	100.0	100.0
2 350–425	38.5	—	33.9	—	14.2	—	3.6	—	0.6	—	9.2	—	100.0	100.0
3 425–500	33.4	30.0	20.6	70.0	21.1	—	1.1	—	0.8	—	23.0	—	100.0	100.0
4 500–600	35.4	36.8	28.1	33.5	24.2	—	0.7	29.7	0.8	—	11.0	—	100.0	100.0
5 600–750	30.6	13.4	16.6	6.0	33.9	57.5	0.6	7.4	0.8	15.7	17.7	—	100.0	100.0
6 750–950	26.8	13.6	19.7	18.9	26.7	42.9	6.1	8.5	1.3	16.2	19.4	—	100.0	100.0
7 950–1250	26.9	14.8	17.5	15.5	31.7	43.8	1.1	14.3	5.7	11.5	17.0	0.1	100.0	100.0
8 1250–1600	24.2	25.7	15.4	21.2	26.4	26.7	2.0	3.3	5.8	21.4	26.2	1.8	100.0	100.0
9 1600–2200	19.9	5.8	7.7	22.1	42.8	28.1	—	3.9	8.2	40.0	21.3	0.2	100.0	100.0
10 2200 or over	18.8	5.7	5.7	24.7	24.3	12.4	—	2.4	46.9	54.8	4.3	—	100.0	100.0
All classes	28.9	16.2	19.2	19.4	27.9	34.3	1.8	7.2	3.9	22.8	18.3	0.1	100.0	100.0

steel. In the PCE ranges above Rs.1600 stainless steel is the major metal in the household stock, aluminium and brass ranking second and third, respectively.

Current purchases

As mentioned earlier, in analysing consumption behaviour of households in respect of consumer durables like metal utensils, the stock is a more appropriate variable than current expenditure, because the latter constitutes only a small proportion of the total household requirements and consumption. On the basis of our survey, the percentage of current purchases (annual) to total value of metal utensils stock with households is estimated at 6 in the case of rural and 4.2 in the case of urban households. But for the purposes of analysing the pattern of demand for metal utensils, particularly with a view to assessing the impact of changes in levels and structure of incomes, the pattern of current purchases would be more useful. Moreover, certain hypotheses could be advanced about the households' expenditure behaviour towards additions to their stock of metal utensils. It could be argued, for example, that the lowest-income groups would add to their stock mainly to increase the number of utensils and to meet their more essential unfulfilled needs, while the higher-income groups may add primarily to their stock in order to introduce variety and qualitative improvements.

Even though the current purchases of metal utensils form a small item in total household expenditure, its quantity and characteristics reflect the nature and magnitude of demand for metal utensils as an item of consumption. It is, therefore, worth while to analyse the behaviour of expenditure on metal utensils among households with different characteristics. Theoretically, the household demand structure is determined by real income levels, family size and composition, individual preferences and relative prices. In the case of metal utensils the size of family seems to be relevant in determining the number of utensils, particularly of the 'serving' category. Social surroundings and differences in the levels of education may influence demand operating through value orientations. Lastly, the demand for an individual item is related to the entire structure of household consumption which is determined by purchasing power, preference structure and relative prices.

A few characteristics of the data on various variables for different PCE groups may be noted (see Table 4.3). The absolute amount of expenditure on purchase of metal utensils shows almost proportionate increase with PCE levels in the case of rural households; but the rate of

Table 4.3 Current demand for metal utensils in relation to household characteristics

Per capita expenditure (Rs./year)	Per household expenditure on metal utensils (Rs./year)		Per capita expenditure on metal utensils (Rs./year)		% of expenditure on metal utensils			
					Total		Cooking and cooking accessories	
	Rural	Urban	Rural	Urban	Rural	Urban	Rural	Urban
Below 350	3.00	3.30	0.50	0.51	0.16	0.19	0.13	0.19
350–425	3.81	–	0.55	–	0.14	–	0.09	–
425–500	5.43	5.88	0.84	1.01	0.18	0.21	0.12	0.21
500–600	6.35	5.77	1.13	1.33	0.19	0.23	0.10	0.12
600–750	7.79	8.39	1.41	1.37	0.21	0.19	0.10	0.07
750–950	11.00	11.65	2.07	2.52	0.24	0.30	0.13	0.05
950–1250	16.91	18.01	3.32	3.56	0.30	0.32	0.12	0.11
1250–1600	25.74	26.08	5.61	5.38	0.39	0.37	0.08	0.10
1600–2200	40.62	55.06	5.93	7.93	0.28	0.36	0.19	0.34
2200 or over	85.33	38.32	18.29	9.23	0.49	0.24	0.08	0.01
All classes	12.53	20.34	2.21	3.91	0.28	0.32	0.14	0.13

increase is faster than PCE in the higher-income ranges of urban households. This would be true both of per household and per capita expenditure on metal utensils as the household size does not vary very much among the different PCE ranges. The increase in percentage expenditure on metal utensils, with the rising levels of per capita expenditure, is however found to be rather slow and much less than proportionate to the variations in PCE ranges. The percentage of expenditure on utensils in total expenditure varies between 0.14 and 0.49 in the case of rural households and between 0.19 and 0.37 in the case of urban households.

It is interesting and pertinent to note here that the expenditure on cooking utensils and accessories reveals quite a different behaviour from the total expenditure on metal utensils. While the latter, as percentage to total household expenditure, increases with PCE levels, the proportion that goes to cooking utensils and accessories shows a declining trend. For example, in the case of rural households, 80 per cent of the expenditure on metal utensils is accounted for by the cooking utensils and accessories in the lowest PCE range of up to Rs.350; the proportion is between two-thirds and one-half in the next four PCE ranges, then steeply declines to less than half reaching a mere 17 per cent in the highest range, with the exception of the PCE range Rs.1600–2200. In the case of urban households, similar behaviour is observed with greater sharpness. In the PCE ranges up to Rs.500 all the current expenditure on metal utensils goes to the purchase of cooking utensils and accessories; in the next range this category claims around half the expenditure on metal utensils but in the PCE ranges beyond Rs.600, its share is very much lower, declining to a mere 4 per cent in the highest range. The PCE range Rs.1600–2200 proves an exception, as for rural areas.

The relation of expenditure on cooking utensils and accessories to total household expenditure does not only show an over-all declining tendency with increasing PCE ranges, but also reveals a sharp discontinuity in the rural PCE range of Rs.600–750 and in the urban class of Rs.750–950. From the bottom PCE range up to these levels, the percentage expenditure on cooking utensils and accessories continuously declines; then it sharply increases in the next PCE range, but again shows a continuously declining trend with the exception of PCE range Rs.1600–2200 in both the rural and urban case. It may be noted here that the dividing line between the poor and non-poor households as identified in our analysis falls in PCE ranges, Rs.600–750 for rural and Rs.750–950 for urban areas. It is, therefore, plausible to suggest that beyond this level of per capita expenditure a definite qualitative change seems to take place in the composition of cooking utensils purchased, in favour of costlier metals and varieties.

The pattern of current purchases of metal utensils by rural and urban households by and large supports the findings of our earlier analysis of the stock of metal utensils. First, the low-income groups both in the rural and urban areas are able to spend a very small amount on metal utensils currently, but most of what they spend goes towards the purchase of cooking utensils and cooking accessories. In the middle-income ranges the major part of current expenditure is shared equally by cooking and serving utensils. In the high-expenditure ranges serving utensils score over cooking ones, though carrying and storage function also claims a significant portion of current expenditure.

What is further significant to note is that a higher expenditure on metal utensils by the households in the upper PCE ranges represents not so much a larger quantity but a higher value per item, while in the lower ranges, say up to Rs.1250 PCE, variations in value mainly represent the number of items purchased by households. This pattern is more particularly discernible in the case of purchases of serving utensils, where one item purchased by rural households in the PCE range Rs.500–750 is valued at Rs.3.69, while two-thirds of an item purchased by households in PCE range of Rs.1200–2200 has a value of Rs.12.50. A similar pattern is discernible in the case of current purchases of urban households.

In terms of the material base, brass, aluminium and iron along with 'others' account for almost the entire expenditure on utensils, in the case of rural households. *Phool* utensils do not seem to be a popular item of current purchases. The expenditure on iron utensils by rural households is almost constant in all PCE ranges, that on aluminium utensils shows an increasing trend with high PCE ranges, expenditure on brass utensils increases slowly up to the PCE range of Rs.1250–1600, but then suddenly shoots up very steeply after that. Stainless steel does not claim a significant part of current expenditure on metal utensils except in the PCE range of Rs.2200 and over.

In the case of urban households, aluminium, iron and stainless-steel utensils have an almost equal share in current purchases. Aluminium and iron utensils claim more or less similar absolute amounts of expenditure among the households in different income ranges. Brass utensils become an item of current purchase only for the households with a PCE above Rs.750, and then the amount spent on this item increases more or less in proportion to per capita expenditure. Similarly, stainless-steel utensils are only bought by households having a per capita expenditure level of more than Rs.750, and beyond this point purchases increase rather sharply.

Annual expenditure on metal utensils per household amounts to Rs.12.5 for rural and Rs.20.3 for urban households. The metal-wise percentage distribution of the value of purchase is: iron 11.4, aluminium 26.5, brass 26.9, stainless steel 2.8 and other metals 32.5 in the case of rural households. The corresponding figures for urban households are 17, 21.1, 17.7, 15.8 and 26.0. It may be noted that the share of stainless-steel utensils markedly differs between rural and urban households. A difference of this order in the purchase of stainless steel, the costliest among the metals considered, is indicative of marked differences between living standards and life-styles of the people belonging to rural and urban areas. Further, the lowest magnitude of expenditure on *phool* reveals the lowest order of preference for such utensils.

4.2 Basic Needs and Metal Utensils

The behaviour of expenditure on metal utensils as revealed by the analysis in the preceding section suggests, on the one hand, that metal utensils do not constitute an item of basic need in the same category as food; but, on the other hand, they show more basic-need characteristics than most other non-food items.

It may be pointed out that cooking utensils show the characteristics of a basic-need item more significantly than the metal utensils group as a whole. This is shown both by common observation and the pattern of expenditure. The pattern of responses from the sample households in respect of the ranking order of various metal utensils confirm this: in the event of a general reduction in utensil prices, an item of cooking category receives the first priority among the lower-income groups, but as we advance higher in the PCE ranges serving utensils claim the priority in most cases. This variation is more marked in the urban than in the rural sample. It may further be noted that 75 per cent of those rural households who would buy additional utensils in the event of an income rise or general price reduction would like to add to their stock utensils of the same quality as they now possess. In this respect, the urban consumers, however, seem a little more conscious of quality because about three-fourths of the prospective buyers would like to have additional utensils either partly or wholly of better quality. In terms of metal categories, the rural consumers would mostly buy aluminium utensils, brass receiving second preference and iron running a poor third. In the urban case stainless steel and brass would share the top priorities.

Identification of the 'basic-need' component of metal utensils

The various dimensions that are necessary to identify the basic-need component of the metal utensils stock possessed by households are: number, uses, metal base and other quality attributes. One way of identifying it could be purely normative, as is done in the case of food. One could specify the minimum number and size of cooking utensils and accessories, number of serving and carrying utensils needed by a household of a given size, in order to enable it to fulfil its minimum needs. The specifics of location, social environment, etc., can also be taken into account in this approach, by empirical observation. There are, however, questions of metal composition of utensil stock, durability, etc., which are difficult to tackle in a purely normative approach; for the same needs could be fulfilled by the use of utensils made of different metal and of different durabilities.

Another approach that can be adopted is that of empirically identifying the basic-need component of metal utensils by comparing level and composition of utensil stock purchased by households who are considered to be just at the level of fulfilling their basic needs in respect of some other more important items. Given the fact that metal utensils comprise essential goods used primarily for preparing and serving food, it can be assumed that households capable of fulfilling their basic food needs would also possess the minimally required stock of utensils. The problem with this approach is then to identify groups of people who – on the basis of food norms, measured against caloric and nutritional norms – are just able to satisfy their basic food needs.

The Indian Planning Commission has estimated that on average a per capita total expenditure of Rs.53 per month in the rural area and Rs.62 per month in urban areas (both at 1973–4 prices) corresponds to the fulfilment of bare minimum caloric intake requirements. The corresponding per capita annual estimates for 1979–80, the year of the survey, amount to Rs.817 for rural and Rs.1161 for urban areas. Taking this as the dividing line, we find that 62.7 per cent of the rural population and 52.8 per cent of the urban population constitute the deprived group (referred to as segment 1 onwards). Thus the non-deprived group (segment 2) represents not even half of the total population as revealed by the survey data. To identify the level and composition of metal utensils for fulfilling basic needs, we have used the

above poverty line estimates. We assume that the availability of metal
utensils which are regularly used by the group of people just around the
per capita expenditure level of Rs.817 per year in rural, and Rs.1161 in
urban, areas represents the minimum absolute requirement. We therefore consider the regularly used stock in possession of the average rural
household in the PCE class Rs.750–950 and urban households in
Rs.950–1250 as representing the requirement of metal utensils fulfilling
the basic need. The average family size of the rural and urban
households in the above PCE ranges are 5.3 and 5.1 persons respectively,
which are not significantly different from the averages for the sample.
The value of stock of regularly used metal utensils in possession of the
rural households in the relevant PCE ranges amounts to Rs.197 and of
the urban households to Rs.416. Some characteristics of these stocks
that satisfy the basic need of household utensils are shown in Table 4.4.

It shows that a rural family of five persons would need four or five
cooking utensils like *tawa*, *karhai*, *batuli* and *bhagona*; three cooking
accessories such as cooking spoons, *sansi* and *chimta*; seven or eight
serving utensils like *thali*, plate, bowls and tumbler; one or two spoons
and around four utensils for storage and carrying purposes. The
minimum required numbers for an average five-person urban family
would be a set of six utensils for cooking, four cooking accessories,
thirteen serving utensils and four for storage and carrying purposes.

TABLE 4.4 Characteristics of basic-need composition of metal utensils

	Rural		Urban	
	Number	Value (Rs.)	Number	Value (Rs.)
1 Stock	19.18	197.34	29.71	415.61
2 Use categories:				
Cooking	4.72	56.66	6.28	134.45
Cooking accessories	3.16	8.54	3.92	18.52
Serving	7.54	77.92	12.90	197.22
Serving accessories	0.66	0.79	3.16	5.50
Carrying and storage	3.30	53.43	3.48	59.92
3 Metal composition (% of value of stock)				
Iron		26.81		14.82
Brass		26.73		43.80
Aluminium		19.65		15.52
Phool		6.14		14.28
Stainless steel		1.26		11.49
Others		19.41		0.09

It may be noted that the minimum requirements of utensils (in numbers) for urban households appear to be higher than for their rural counterparts. These findings are in line with the observation that the needs, perceptions and life-styles of the rural and urban people are different. The number of food items cooked and served, the practice of taking meals jointly or separately or at different hours during a day, and other differences between rural and urban people, seem significant in determining the household requirements. When asked directly whether they feel that their stock of metal utensils is adequate to meet their minimum need, 57 per cent of the urban consumers gave a negative answer, while of the rural consumers only one-third considered their stock inadequate.

4.3 Supply Conditions: Technology and Marketing

For the analysis of technology we have used the metal categories of utensils, because preliminary observations showed that inter-metal variations were more significant than within-metal differences. This was confirmed by the analysis of the twenty-three sampled units which were distributed as follows: five units each producing iron (Kanpur), aluminium (Lucknow), brass (Moradabad) and *phool* (Basti), and three units producing stainless steel (Ghaziabad). If variations in size and use of total productive capital (fixed *plus* working capital) per worker are used as indicators of technological levels, it is seen that while each of the metal groups stand out clearly in distinction from others, all or most units within a metal group lie within a relatively narrow range of these variables.

Products, materials and equipments

Major items produced by the sample metal utensil manufacturing units are *karhai*, *tawa*, *tasla* and can (bucket), each of iron; *batuli*, *pateeli* (aluminium and *phool*), kettle (aluminium), tumbler (aluminium, brass, *phool* and stainless steel), *parat* (aluminium, *phool*) and bowls (aluminium and stainless steel). The sample stainless-steel units also manufactured certain other items, such as plates of different sizes, *bhagonas*, rice trays, compartmental trays and cafeteria trays.

The iron utensils are produced from iron sheets, available in different thicknesses, at the government-controlled rates, through a quota system and also from the open market. The aluminium units used aluminium scrap and ingots for producing the utensils. The brass and *phool* industry mainly recycled old material in the form of rejected utensils and scraps. Unlike these, the stainless-steel units used imported sheets to a large extent. Other major material inputs used comprised steam coal, hard coal and fire wood particularly in brass and *phool* industry, rivets in iron industry, in addition to power, chemicals and paints.

For production of iron utensils like *karhai*, *tawa* and *tasla*, mostly hand-operated tools are used. They are rail (a piece of railway line used as base for hammering), *nai* (anvil, an iron based instrument used for shaping utensils by hammering), *chheni* (a hard iron rod flattened and sharpened at one end, used for cutting by hammering), *nihai* (a bigger size *chheni* with wooden top), hammer, *ghan* (a large-sized hammer) and scissors. For producing cans, power-operated equipment such as cutting machines, roller and core machines are also used.

The aluminium utensils industry uses more mechanised and sophisticated equipments like rolling machine, cutting machine, press machine, lathes, electric drills, shapers, polishing equipments, etc., besides electric motors, hand tools and consumable stores.

The brass units deployed lathes, scrapping machine, rolling machine, cutting machine, punching machine and buffing machine. The punching machine is mainly used for producing cans. The *phool* industry mainly uses hand-operated tools and equipment such as melting pot, fan, furnace, frame, pincers, hammers and anvil. In the manufacturing of tumbler, bowl and *parat*, turnery and polishing machines operated by electric power or diesel are also used. The steel units deployed press machine, cutting machine, spinning lathe, shaping machine and polishing machine.

Processes of production

The items like *tawa*, *karhai* and *tasla* are produced from iron sheets that are cut into circular discs and then hammered for shaping with the help of anvil and hammer. The distinction among *tawa*, *karhai* and *tasla* is mainly of size and depth. In *karhai* two loop-shaped handles are also fitted. The handles are made by bending iron rods, punching and fitting them on two opposite sides with the help of rivets. The production of these utensils is done mainly with the help of hand-operated tools. The manufacturing of iron cans is done with the help of power-operated equipment, used for cutting iron sheets, shaping, making side rings in which the handles are hinged, and for punching. The hand-operated tools are also used for bending, shaping and fixing rivets. The cans are also polished after checking for any leakages.

Manufacturing of aluminium utensils involves melting, rolling, disc cutting, pressing, spinning, washing and polishing processes. Melting and rolling become necessary when scraps and ingots are used for producing sheets. Aluminium sheets are also available for production of utensils. The sheets are cut into discs and pressed or spun for shaping. The utensils are then washed, polished and packed.

Among the brass utensils, the manufacturing of items like tumblers involves melting, moulding, scraping, smoothing and polishing. The moulding is done with the help of a furnace, such that melted metal drops into the moulds and the unfinished product is ready. Then follows scraping of the utensil with the help of diesel or electric-powered machines. The smoothing and polishing processes are carried out with the help of lathe and buffing machines. Other brass items like *thali* and cans are produced through cutting, pressing, shaping and polishing processes. In the case of cans, a punching machine is also used to make holes for fixing handles.

The manufacturing of *phool* utensils, as indicated earlier, is done mainly with the help of hand-operated tools. The processes involved in the production are melting and rolling, pressing, grinding and polishing for items like *thali* of different sizes, tumbler and bowl. A rolling mill that converts the metal blocks into sheets caters for a number of household units. Only a few units possess powered grinding and polishing equipment. Most of the units carry out these processes manually.

The process of manufacturing stainless-steel utensils begins with cutting the metal sheet into circular (or rectangular) discs as required for specific items. The discs are cleaned, oiled and pressed with the help of a press machine. In this process various kinds of dies are used for shaping. This is followed by a spinning process for bringing in accuracy in the shape, and turning round the edges. The utensils are then polished and packed. In making rectangular items like some of the trays, the spinning process is not involved.

Production, employment and capital-intensity

The total investment per unit in the stainless-steel sector is of the order of Rs.684 000 followed by aluminium at Rs.424 000, iron Rs.84 000, brass Rs.64 000 and *phool* Rs.15 000 only. Except for iron, similar ordering of a more pronounced kind is observed in the value of fixed assets per unit in different metal categories. The per unit value of fixed assets in the iron and *phool* industry is of about the same order since units in both these sectors mostly use hand-operated tools. Yet the value of output per unit in the iron sector is almost four times that in the *phool* sector. It is neither possible nor necessary in the present context to ascertain the contributions of technology and relative prices to the disparity between the output levels of the iron and *phool* utensil manufacturing. Yet, the observation is consistent with the fact that while the demand of iron utensils has been growing, that for *phool* utensils is on the decline because of scarcity and the high price of the metal. The situation has led to a shrinking of *phool* industry.

For every 100 rupees of output, the total cost for *phool* utensils comes to Rs.87.3, which is the highest since total costs for brass and stainless steel stand in the vicinity of Rs.75 and for aluminium at Rs.66.1, the lowest among the five categories. The amount of profit earned per sample unit for stainless steel is about Rs.204 000 followed by aluminium Rs.124 000, iron Rs.75 000 and brass Rs.23 000. Profit as a category of accrued income is not relevant for *phool* units where the surplus mainly represents the notional wages of unpaid family labour. The profitability seen as a percentage of profits to the cost of production ranges between 19 and 23 for all the metal categories. But the amount of annual profit as a percentage of total investment comes to as high as 89 in the case of iron units, against the range 29–35 for other metal categories. It seems that the growth of the iron utensil industry is the least constrained as compared to the units in other metal categories.

The usual number of workers on a particular day per sample unit is highest for aluminium (27.4), followed by stainless steel (14.0), iron (9.4), *phool* (6.0) and brass (5.0) in that order. In the *phool* units hired workers were casually engaged for only a few days during a month for jobs like melting and casting. Thus the number of workers reported for the *phool* units mainly represent household workers. The average numbers of man-days deployed annually per sample unit are estimated at 8220 for aluminium, 4200 for steel and 2920, 1804, 1386 for iron, *phool* and brass, respectively.

The wages in brass, *phool* and iron units are generally paid on a piece-rate basis. In the case of aluminium, three of the five sample units reported to be paying wages mainly on time-rates. All the stainless-steel units reported only time-rates for wage payments. The average monthly wage levels of the workers come to Rs.218 and Rs.217 for aluminium and stainless-steel units, Rs.207 and Rs.202 for brass and iron, and only Rs.190 for workers in the *phool* sector. The comparison between the metal units therefore shows positive relationships between the levels of investment, wages and production technology.

To the extent that capital-intensity measures the level of technology, the stainless-steel sector is at the top, followed by aluminium, brass, iron and *phool* (Table 4.5). This is in conformity with the fact that while stainless steel and aluminium are modern industries, the manufacturing in brass and iron sectors have a mix of traditional and modern technologies and the *phool* industry is carried out mainly on traditional lines. Certain other technical ratios of production also conform to this inter-metal pattern. The proportion of value added to value of output amounts to 32 per cent for aluminium, followed by brass 25, iron 28, stainless steel 21 and *phool* 13. The inter-metal ordering remains unchanged in terms of wages–output ratio as well as share of wages in value added. The employment potential per unit of investment appears to be highest in *phool* industry, followed by the iron, brass, aluminium and stainless-steel industries. This suggests an inverse relationship between employment potential and the level of technology, in so far as only direct employment is considered.

An assessment of the employment potential per unit of investment is relevant particularly in the context of employment planning. To assess this potential, one needs to take into consideration the labour coefficients that are defined as the amount of labour required per unit of production. There is a wide variation in the labour coefficients,

TABLE 4.5 Technical ratios in metal utensil manufacturing

Item	Metal category				
	Iron	Aluminium	Brass	Phool	Stainless steel
Total investment per worker (Rs.)	8 958	15 469	13 046	3 527	48 880
Value added to output ratio (%)	25.05	32.35	25.27	12.71	20.50
Wages paid per Rs.100 of output	5.67	11.55	8.68	1.00	2.84
Wages paid to value-added ratio (%)	22.23	34.35	34.35	7.88	13.85
Annual number of man-days per Rs.100 of investment	3.35	1.94	2.17	12.13	0.61
Number of man-days deployed per Rs.100 of output	0.70	1.12	0.99	1.67	0.33

measured as man-days per Rs.100 of output. Table 4.5 shows figures at only 0.33 for stainless steel; and those for iron, brass, aluminium and *phool* are around two, three, four and five times that of the stainless steel.

The pattern of labour coefficients makes the ordering of metals for utensil manufacture by the criteria of minimum need fulfilment, and employment potential, somewhat difficult, since employment intensity, labour productivity and basic-need characteristics of different metals vary in different orders. Stainless steel has the smallest employment potential per unit of capital, but has highest labour productivity. It is, however, least important for fulfilment of basic needs. *Phool* is the most labour-intensive category of utensils but has the lowest productivity per worker. Its high prices and non-availability are, however, making it increasingly irrelevant for the purposes of meeting basic needs. The remaining three metals – iron, aluminium and brass – have one order on the basis of employment-intensity, but a reverse order on the basis of labour productivity. However, as far as minimum needs fulfilment is concerned, the ordering of the three metals is the same as in the case of employment-intensity; the productivity differences among the three categories of metal utensils are not very wide.

Production and price trends

Let us also look at the production trends in recent years and see how they conform to the requirements of basic-needs fulfilment. Information from the sample production units in different metal categories for the past five years reveals that the pattern of growth hardly corresponds to the pattern of supply required for the fulfilment of basic needs. Of the five metals considered by us, units engaged in the production of stainless steel have grown fastest at an average annual growth rate of output of around 20 per cent. Aluminium utensil units came next with an average growth rate of 9.5 per cent per year, iron units are growing slowly at around 5 per cent per year, while production of brass units has stagnated for a few years; growing brass units are mostly engaged in the artware production mainly for export. Production of *phool* utensils is facing actual decline. To a certain extent these trends correspond with the pattern of demand emerging out of the trends in the distribution of incomes: but they also reflect, to a similar degree, the production constraints mainly in the form of the supply of raw material. Most units, particularly those engaged in the production of *phool*, brass and aluminium utensils, attribute their slow growth or stagnation to the non-availability of raw material.

It is, however, not the absolute non-availability of metal utensils, but low levels of incomes and high and rising prices, that have posed the major constraints in the fulfilment of basic-needs requirements of the households. No respondent household indicated non-availability in general or in the local market as a factor responsible for their non-fulfilment of requirements or for the absence of metal utensils as an item of current purchase. Of about 40 per cent of the rural and 55 per cent of the urban households that considered their stock of metal utensils short of their basic requirements, low income was mentioned as the reason by 80 per cent of the rural and 55 per cent of the urban households; the rest ascribed their inability to meet their requirements to the high prices of metal utensils. In effect, the two reasons could be identified as a single factor: lack of purchasing power. Metal utensils prices are found to have increased fast during the period 1965–80 (Table 4.6): the fastest increase of over 300 per cent has been registered in the prices of iron utensils, followed by stainless steel (245 per cent), *phool* (200 per cent), brass (166 per cent) and aluminium (130 per cent). The general rise in money incomes during the same period has lagged far behind. Consequently, a larger proportion of potential buyers of metal utensils do not get turned into effective buyers. Our investigations into this aspect from the traders reveal that around 61 per cent of the customers who visit their stores go away without making any purchases, after inquiring about prices. The proportion of such customers is found to be similar in the case of both urban and rural retailers.

Income as well as price constraints are found to be more severely felt by the rural than the urban consumers. Due to a generally low population income index of rural areas, trading in metal utensils is not a very competitive activity and also involves a larger risk element owing not only to small magnitude but also of irregularity of the demand for

TABLE 4.6   Retail price (per kg) of metal utensils (1965 and 1980)

Metal category	Rural outlets		Urban outlets	
	1965	*1980*	*1965*	*1980*
Iron	2.00	7.50	1.50	6.00
Aluminium				
Branded	25.00	40.00	15.00	32.00
Unbranded	12.50	32.03	10.00	24.54
Brass	15.00	40.63	12.00	32.00
Phool	22.00	69.96	20.00	60.00
Stainless steel	40.00	135.00	35.00	120.00

metal utensils. Consequently, the rural outlets expect a higher margin of profit on their sales. But the constraint of local purchasing power puts a check on their margin of profit. One of the methods adopted by them to tackle this problem, as was observed during our field investigations, is to sell low-quality items at the usual price for the items of standard quality. The unbranded nature of most of the items sold, gullibility of rural buyer and lack of competition in the area easily enable them to adopt such a practice. It is not uncommon to come across the sale of unbranded aluminium and low-quality iron and brass utensils at the prices of branded and better-quality items in rural areas. It has not been possible to quantify empirically the extent of such practice, for obvious reasons, but the practice is found to be commonly prevalent.

Marketing channels and rural–urban price differences

Despite this 'facility' of manipulation by the rural retailers, the average prices paid by the rural consumers, irrespective of the quality of utensils, are substantially higher than the urban prices, as can be seen from the data collected by us from the rural retailers (Table 4.6). We have attempted a factual analysis of these differences between the rural and urban areas, in terms of the additions to price over the ex-factory price, by way of transport cost, taxes, octroi and traders' mark-up, at each sale point in the trade channel. Let us first take taxes. There is the production tax called excise duty; it is levied at the rolling stage of metal processing and gets included in the price of raw material. No excise duty is levied in the subsequent stages of manufacturing of metal utensils as such. The production tax thus makes a common part of the cost irrespective of the mode and area of sales of metal utensils. Then there are central and state sales taxes. The central sales tax is levied at the rate of 4 per cent *ad valorem* on utensils imported from outside the state. This again would not account for rural–urban price differences as there is no indication that utensils from outside the state are mostly sold in the rural areas. In fact, it is only the stainless-steel utensils that are usually subject to this levy and they are mostly consumed in the urban areas. State sales tax is a single point tax levied at a rate of 8 per cent *ad valorem*. Onus of collection of this tax is on the first seller, that is, the manufacturer, and is, therefore, included in the manufacturers' price. All these taxes would thus not make a difference in the price of metal utensils between the rural and urban areas.

Taxes and rates levied on movement of goods would, however, make a difference depending on how many levy points the goods pass. Goods manufactured and sold in the same town will not be subject to octroi duty, which is levied on goods entering into certain specified areas, mainly the city municipal limits. The general rate of octroi is reported to be Rs.5.00 per quintal for iron, aluminium, brass, copper and *phool*, and Rs.60.00 per quintal for stainless-steel utensils. There is reason to believe that the utensils sold in rural areas would generally be subject to at least one more levy of octroi than those sold in urban areas if the manufacture of metal utensils is not located in the town from where the rural retailers procure their supplies. It would thus account for some difference in the prices between the urban and rural areas. But it forms a small part of the total price and thus could not account for a very significant part of the rural–urban differences in prices. This is also found to be the case with transport charges, the major part of which is incurred in the first stage, that is, transport of goods from the place of their manufacture.

The urban areas are found well serviced in terms of the number of retail sale outlets but, for the reasons stated earlier, rural retail outlets in metal utensils are few and far flung. The regular village retailers, who sell utensils along with other items, are supplemented by the weekly markets that have a few stalls selling utensils. The following features of the marketing of metal utensils are interesting and useful, particularly with a view to analysing the difference between the urban and rural retail prices:

1. The major part of the purchases by urban retailers is made directly from the manufacturers, either locally or situated in other towns. The rest who buy from the wholesalers, mostly local, have to pay higher prices as it also includes wholesalers' margin.

2. All rural retailers report their purchases from urban wholesalers and their purchase price should generally be the same as that of the urban retailers. Closer scrutiny, however, reveals that the dealers from whom the rural retailers buy are wholesalers-cum-retailers. On smaller purchases, such as by rural retailers, the price charged by them is not the wholesale but retail price. The proportions of rural retailers' purchases at the wholesale and retail prices are roughly estimated to be equal. All urban retailers, on the other hand, buy from wholesalers at the wholesale price.

3. Transport and octroi charges are not significantly different among the routes of procurement, and in any case the somewhat higher charges in the case of rural retailers cannot justify the magnitude of price difference between rural and urban areas, as transport and octroi make only about 1 to 2 per cent of the total purchase price.

Thus the major reason for price differences between rural and urban areas is to be found in the route of procurement by retailers and the purchase price they have to pay for metal utensils. We have attempted an estimate of the effective cost of procurement of the metal utensils by retailers by different channels. We have taken the manufacturer's sale price as 100 and worked out the cost to the retailers by adding transport costs, octroi and trade margins of the intermediate sellers. We find that the metal utensils which the manufacturer sells at Rs.100, inclusive of excise and sales tax and his own profit, get ultimately into the hands of the urban retailer, on an average, at a cost of Rs.111 and of the rural retailer at a cost of Rs.130 (Table 4.7). A small part of Rs.1 to Rs.2 of this difference is the result of transport cost and octroi, but most of it is accounted for by the different sources of procurement and different

TABLE 4.7 Effective purchase cost of retailer per Rs. 100-worth of utensils at manufacturer's sales price (including excise and sales taxes)

Urban retailers

| Metal | Source of procurement by retailers | | | | | |
| | Manufacturer | | Wholesaler | | Average cost of purchase | Average sale price |
	% purchase	Cost (Rs.)	% purchase	Cost (Rs.)	(Rs.)	(Rs.)
Iron	50	104	50	123	114	135
Aluminium	69	102	31	118	107	125
Brass	80	106	20	125	110	130
Phool	20	103	80	118	115	131
Stainless steel	80	105	20	122	108	133
Average	60	104	40	123	111	132

Rural retailers

| Metal | Wholesalers at wholesale price | | Wholesalers-cum retailer at retail price | | Average cost of purchase | Average sale price |
	%	Cost (Rs.)	%	Cost (Rs.)	(Rs.)	(Rs.)
Iron	60	124	40	135	130	147
Aluminium	50	119	50	127	123	148
Brass	30	127	70	138	134	158
Phool	55	119	45	133	126	137
Stainless steel	60	123	40	135	127	152
Others	35	123	65	135	131	148
Average	49	124	51	135	130	149

prices paid by the retailers. A retailer gets the utensils cheapest if he procures directly from the manufacturer: over the price charged by the latter he only incurs an additional cost of Rs.4 on Rs.100 worth of utensils. A retailer purchasing from the wholesaler in addition pays the latter's margin so that he gets utensils worth Rs.100 at manufacturers' price, at Rs.123. In the urban areas, 60 per cent of the retailers procure utensils directly from the manufacturer and 40 per cent from the wholesaler. The average cost to the urban retailers per Rs.100 worth of utensils at manufacturers' price comes to Rs.111.

Rural retailers buy from the urban wholesalers, but in about 50 per cent of cases the prices charged to them by wholesalers, who are also retailers, are urban *retail* prices. The effective average cost to the rural retailer, including transport cost and octroi, allowing for Rs.100 worth of utensils at manufacturers' price, comes to Rs.130. The final prices charged by the rural retailers to the consumers, for different categories of metal utensils, are higher in rural areas than in urban areas, more or less in proportion to the differences in cost of procurement by the urban and rural retailers.

Existing marketing arrangements thus operate to the disadvantage of the rural consumers. To a certain extent the small magnitude, low density and irregularity of demand in the rural areas make this disadvantage inevitable. Continuation of this disadvantage with the generally rising trend in prices may increase the prices of utensils so high in the rural areas that an increase in incomes may not lead to a rise in the demand for metal utensils, as envisaged in the subsequent analysis in this study. Establishment of public or co-operative wholesale outlets in the urban areas to ensure that rural retailers are charged wholesale and not retail prices should, however, help reduce this disadvantage.

4.4 Impact of Income Redistribution and Growth

The impact of various scenarios of redistribution and growth is estimated with a static simulation model, which includes labour, technical and consumption coefficients derived from the surveys.[2]

To estimate the expenditure elasticities, the linear hyperbolic, semi-log, double log and log inverse forms of Engel functions were used on household-wise PCE data for expenditure on metal utensils and the PCE level. Based on the statistical criteria such as the maximum value of the coefficient of multiple determination (R^2), minimum value of the standard error of the estimate (s.e.), significance of the statistic 't' and the minimum value of the constant term 'a' – along with the plausibility of economic interpretation, the double-log function appears to be the most appropriate among the given alternative forms. The expenditure elasticity of demand for metal utensils has, therefore, been adopted to be 1.429 for rural and 1.358 for urban households. Given the high value of expenditure elasticity of demand for metal utensils – 1.429 for the rural and 1.358 for urban areas, on the one hand, and invariability of labour and capital coefficients across different simulations, on the other – the effect on demand, capital and labour requirements vary among different simulation situations in proportion with the respective growth in PCE stipulated in each simulation. The other and more important point to be noted is that all increase in demand and other variables is accounted for by increase in PCE level; redistribution is, in fact, found to produce a negative effect on total demand and therefore also on capital and labour requirements. In other words, the effect would have been higher if simulations did not envisage any redistribution at all. What in effect it implies is that since a major part of the demand for metal utensils in value terms comes from the higher-income groups, the simulation that envisages not only a high income growth in general, but of those above poverty line in particular, would lead to the largest positive effect on the demand and other related variables. If the growth in income has a large redistributive component, in favour of the poorer groups, even a rise in average income level of those groups would produce a smaller total effect on demand for metal utensils and employment in their production. By implication, a pure redistribution of income from the non-poor to the poor group would lead to a decline in effective demand.

4.5 Conclusions

On an *a priori* basis, metal utensils can be considered as an essential item of household consumption. It is possible to define, either on an empirical and/or a normative basis, the minimum level of utensil stock to fulfil the basic-needs requirements of cooking, serving and storage. The expenditure on metal utensils, both in absolute terms and as a percentage of total expenditure, is found to increase more than proportionately with the rise in total household and per capita expenditure. To a certain extent this is accounted for by the larger number of utensils, particularly of the serving category, purchased by households at high PCE levels, but, to a large extent, also by the switch to superior quality of the utensils with increasing income levels. As a result, not all purchases of metal utensils can be considered as satisfying basic needs. The basic criterion on which quality difference could be gauged consists in the metal base of utensils which also corresponds to the relative prices per unit of weight. Iron, aluminium, brass, *phool* and stainless steel – in that order – are higher-quality metals.

It has not been possible to analyse the consumption behaviour in respect of metal utensils in disaggregated categories of use and metal base. But considerable empirical and descriptive evidence has been thrown up by our study to show that cooking utensils and iron and aluminium utensils come quite close to qualifying as basic-need items. And it is these that make the major part of the stock and current purchases of the people in PCE ranges below what can be considered a level of minimum need fulfilment. While the value of total stock as well as current purchases of metal utensils show a continuously increasing percentage of total household expenditure, the value of stock and current purchases of cooking utensils and of iron and aluminium utensils decline as a percentage of household expenditure and that of metal utensils.

Cooking utensils and a minimum stock of serving and storage utensils, mainly made of iron and aluminium, would thus make an item of basic need. They do not show the same degree of basic-need characteristics as food items, but certainly rank higher than most non-food items. Using the latest available estimates of the income level identified as cut-off point for poverty estimates, we find that the households with a per capita expenditure level which is presumed to be just enough to meet the minimum basic needs have a stock of metal utensils worth Rs.197 in the rural and Rs.416 in the urban areas. This then is taken to be the minimum value of the metal utensils which fulfil basic needs. The number of items by use category and metal composition of the estimated basic-need basket of metal utensils was also identified. There is considerable evidence to suggest that the households below and above this cut-off point purchase metal utensils of quite different quality – in terms of use category, metal category and prices.

We then looked into the process and economics of production in order to identify supply and technology constraints on production and to assess the relative capital-intensity and employment potential. This analysis was carried out on the basis of inter-metal comparison. Availability of material is found to be posing a serious constraint in the production of brass and *phool* utensils. Such constraints do not exist as much in the case of other metals. However, the demand pattern that emerges from the distribution of growing incomes is going to mainly favour stainless steel. As a result, the production of stainless-steel units has grown fastest; despite high prices aluminium units rank second and iron units third in terms of growth of output in the recent past.

It is, however, not the absolute non-availability of metal utensils, but their rising prices and low incomes, that act as a constraint on their consumption up to the level of fulfilment of basic requirements. Steep price rises have posed this problem for all consumers but more particularly for those in the low-income groups. In addition, the prevalent marketing practices seem to put the low-income households and the rural consumers at a distinct disadvantage. It is seen that the mark-up in consumer prices over the manufacturers' price is higher for utensils consumed by the poor than for those consumed by the non-poor. Brass and stainless-steel utensils that have little relevance for the fulfilment of basic needs are purchased by retailers directly from the manufacturers and therefore have only two margins of profit added to the cost: the manufacturers' and the retailers' margin. Iron and aluminium utensils more often pass through wholesalers, thus adding one more margin to the consumer price. Secondly, the utensils purchased by rural consumers mostly have the manufacturers', wholesalers' and retailers' margin added to the prices, whereas the wholesalers' margin is absent in the majority of cases of the urban prices. This leads to a substantial difference between the urban and rural consumer prices of metal utensils, and to a significant disadvantage to the rural consumers. Thus the prevalent marketing and distribution system of metal utensils militates generally against the supply of metal utensils for the fulfilment of basic needs of the poorer groups of the population.

Inter-metal differences in technology of production of utensils are significant so far as capital and labour coefficients are concerned. The production of stainless steel, aluminium, brass and iron and *phool* requires less capital per worker, in this order. The labour requirements per unit of output are the lowest in stainless steel, followed by iron, brass, aluminium and *phool* in this order. Differences are large between stainless steel on the one hand and *phool* on the other: employment per Rs.100 of capital amounts to twelve man-days in *phool* and only 0.6 man-days in stainless steel, and per Rs.100 of output to 1.7 man-days in *phool* and 0.3 in stainless steel. Differences among the other three metals are not so large, though quite significant. Man-days of employment per Rs.100 of capital ranges between 1.94 in aluminium and 3.35 in iron: and per Rs.100 of output between 0.70 in iron and 1.12 in aluminium. The metals that are mostly used for the basic-needs group of utensils – iron and aluminium – thus have reasonably high labour intensity, but a lower productivity than stainless steel.

These technological differences, however, do not lead to significant differences in the over-all technology and employment situation among the various simulations on growth and distribution paths. Thus, despite considerable inter-metal differences in technology, the over-all technological ratios in the metal utensils production are not very much different in the simulated from the observed one.

Given the relative invariability of the aggregate technical coefficients across various simulations, the total change in the volume of demand for metal utensils and employment generated in their production is solely dependent on the change in levels of per capita expenditure. A higher than unit elasticity of expenditure, which was found for metal utensils, leads to a change in demand and employment generation proportionately higher than the change in per capita expenditure level. But a redistribution of income in favour of those below the poverty line is found to have a negative impact on the total demand and employment generation in the metal utensils sector. This is because a larger proportion of a given amount of incomes in the hands of the relatively higher-income groups is likely to be spent on metal utensils than if the same income is distributed among the poor. This is so because the rich buy qualitatively superior and, therefore, more valuable utensils; and because the poor have other needs to fulfil with their additional income. The conclusion is therefore somewhat perplexing from the viewpoints of poverty alleviation, fulfilment of basic needs and creation of employment. In fact, redistribution of income has a negative impact on the production and the creation of employment in the metal utensils sector. However, growth in incomes would lead to an increase in demand for metal utensils, more particularly from the higher-income groups; increase in production would increase more or less according to the existing pattern if raw materials are available; and employment would increase more or less in proportion to the increase in total demand. It is likely that a somewhat increasing proportion of utensils with higher non-basic need elements would characterise the pattern of incremental demand and production. Even a redistribution of income in favour of the poor, either directly or through employment creation, would not lead to a shift in demand and production in favour of basic-needs fulfilment or in favour of higher labour intensity and greater employment in the production of metal utensils. In fact, such redistribution is likely to lead to a slower rate of growth in demand, production and employment, given the growth rate of incomes.

Notes

1. Dr Papola and Mr Sinha are respectively director and research collaborator at the Giri Institute of Development Studies, Lucknow (India).
2. An explanation of the model can be found in T. S. Papola and R. C. Sinha, *The Consumption Behaviour and Supply Conditions of Metal Utensils in India: A Study in the Basic Needs Framework* (Geneva: ILO, 1982; mimeographed World Employment Programme research working paper; restricted).

5 Income Distribution, Technology and Employment in the Footwear Industry in Ghana

George A. Aryee[1]

5.1 Introduction

The choice of footwear for this chapter was inspired by two consider-ations: first, footwear is a basic commodity purchased by both poor and rich; and secondly, it is a capital-light product that is often recom-mended as being an ideal industry for developing countries. The study is based on Kumasi, the main centre for footwear production in Ghana, possessing over 50 per cent of the formal sector's footwear production capacity.

The chapter is organised as follows. In the remainder of section 5.1 the differences between formal and informal sector firms are considered from an institutional point of view, and taking account of the implications for technological choice. Section 5.2 summarises the data collected on consumer demand for footwear in Ghana, noting the references of different income classes for different types of footwear. The implications for demand of a redistribution of income are examined, and this analysis is brought to a conclusion in section 5.3 which translates the increments in demand for different types of footwear in terms of employment creation, using coefficients obtained from data-producing enterprises. The key findings of the chapter are briefly summarised in section 5.4.

Previous studies using national input–output tables have failed to establish clear evidence about the potential employment-generation effect of income redistribution (see Chapter 1). A key reason for this is that the input–output data for producing sectors is often aggregated in a way that takes no account of important differences in the technologies applied by firms of different sizes. Small informal sector firms are usually excluded from census data because of their size, or because such firms are unregistered. The present study contains many comparisons of such firms in Ghana with formal sector firms.

Another special feature is that account is taken of the differences between skilled and unskilled labour. This distinction is important in so far as the employment of unskilled labour in informal sector firms may be increased by income redistribution measures to a greater extent than the employment of skilled persons.

Bearing in mind these two considerations, the remainder of this first section introduces the footwear industry in Ghana and outlines its structure, both in the modern sector and in the informal sector. As discussed further below, production units in the shoe industry in Kumasi apply different technologies. Furthermore, earlier studies[2] have shown that small informal firms in the industry tend to use unskilled apprentice labour whereas large firms employ wage labour. It is therefore necessary to disaggregate various kinds of firms to analyse our hypotheses in a comprehensive fashion.

The footwear industry in Ghana

The footwear industry in Ghana includes very modern footwear factories and artisan workshops. Some 2500 persons are employed in the footwear industry in Kumasi. Of this total, small informal-sector firms account for 1800 (72 per cent). The industry employs about 12 000 persons in Ghana as a whole.

The largest footwear factory in the country, in Kumasi, started production in 1967. It was designed to produce 2 million pairs per year and cost N₵6 million to build and equip. It employs some 600 persons with N₵10 000-worth of fixed capital per employee. In 1977 this factory alone accounted for 50 per cent of total capacity and 30 per cent of employment in large-scale footwear manufacturing (thirty or more persons employed therein). Ten other large-scale firms share the remaining capacity and employment. At the beginning of the 1970s, footwear production in the large-scale sector accounted for over 55 per cent of total production of footwear in the country, with medium- and small-scale firms (having thirty or less persons employed) producing the other 45 per cent of footwear.

Although the medium- and small-scale sectors produce less than half the total output of footwear in Ghana, they probably employ four times as many people as the large-scale sector. The 1970 population census revealed that large-scale production units accounted for only 18.6 per cent of total employment in the footwear (including repairs) industry. Thus most footwear workers were employed in relatively small- and medium-scale firms.

The relative shares in employment and output suggest that there is a significant difference in labour productivity in the large- and medium-/small-scale sector, because of differences in production techniques and factor prices and in production functions (to be discussed later).

The footwear industry in Ghana depends heavily on raw materials from abroad for its operations, a fact that is not at first obvious. In 1968 the industry imported about 94 per cent of all raw materials.

There is a small tannery in operation in Kumasi and two firms undertake production of artificial leather in the country. However, their output is so small that substantial quantities of both natural and artificial leather have to be imported. The raw materials used in the production of natural and artificial leather are also imported.

The material usually used for making soles is microcellular rubber sheets. These are partly produced with imported material by a rubber products company in Accra. However, a substantial quantity is imported, often in the form of unit soles. Other items made in Ghana with entirely imported input materials are boxes for packing footwear. These are supplied by the Paper Conversion Division of the Ghana Industrial Holding Corporation (GIHOC). Products like buckles, insole boards and crepe rubber are wholly imported. Its dependence on imported inputs has caused the industry to operate at a very low level of capacity utilisation because of the continuing foreign exchange difficulties since the mid-1970s.

Institutional framework and choice of production technology

There are considerable differences in the 'status' of labour utilised in production units in urban Africa. The types of labour employed depend on the size of the enterprise. In Ghana, the larger firms with more than thirty persons engaged in the shoe and in other manufacturing industries employ labour on a wage basis. However, smaller firms, especially those with less than ten employees in the so-called informal sector, employ different types of labour.

Previous studies on the informal sector in Africa indicate that the type of labour employed in this sector includes the self-employed person, family labour, apprentices and partners. Table 5.1 shows the type of

T_ABLE 5.1   The enterprise

Single owner and single-operated	Single owner	Partnership
Own labour	Own labour	Own labour
	Family labour	Partners
	Apprentice labour	Family labour
	Casual labour	Apprentice labour
		Casual labour

ownership and the status of the 'employees'. Broadly summarising, there are three types of ownership system in the sector: the single-operated enterprise; the single owner but not single-operated enterprise; and the partnership system of ownership.

In all these systems of ownership, the self-employed entrepreneur plays a crucial role in direct production. He also performs entrepreneurial and managerial tasks. There is hardly any division of labour in this respect. Ownership and control of the firm reside in the working owner. The partners are usually ex-apprentices of the enterprise. When an apprentice, after completing his training, is unable to secure a job or establish his own business, he may be allowed to join his ex-master on a profit-sharing basis. This sort of arrangement is usually temporary. The partner stays long enough to save sufficient money to establish his own business or until he gets wage employment. Thus partnerships are rare in the informal sector.

Apprentices account for some 60–70 per cent of all persons engaged in the informal sector. This proportion indicates an important function of the informal sector, namely training.

The relationship between a master and his apprentices differs significantly from that between an employer and his employees in the formal sector. To understand this, it is necessary first to examine the factors affecting self-employment and those determining demand and supply of apprentices.

The fact that each establishment has at least one working owner (the self-employed person) is one of the distinguishing features of the informal-sector establishment. To establish himself in self-employment, the working owner must have a complete knowledge of the trade and must be self-sufficient in production and marketing. Knowledge of the trade is therefore of utmost importance. It may even be regarded as being more important than capital because the individual has to acquire the necessary training in the trade before the need for physical capital arises.

Consequently, the relationship between a working owner and his apprentice is structured by the process of passing on knowledge, not by wage rates. However, apprentices are often given daily meals and occasionally some pocket-money to buy cigarettes. These sums of money may be regarded as bonuses for good work rather than wages. Their value is usually no more than 10–15 per cent of the minimum wage paid to workers in the small-scale sector[3] and the apprentice cannot generally live on it. Moreover, the apprentice pays an entrance and an exit fee to his master. To earn more, the apprentice must work for his own customers in his leisure time or depend on his own family members and relatives. Indeed, for individuals who are not relatives of the master, a condition for entering apprenticeship is that he or she has a 'patron', some personal savings of his own, or is in a position to borrow money to defray the costs of apprenticeship. Thus the apprenticeship system, in various ways, tends to subsidise labour costs to the employer.

The behaviour of the apprentice is akin to that of the family member who, after migrating elsewhere from the family for a job, continues to obtain family economic support. In such a case, he does not 'have to seek compensation for an income loss; the supply price of labour will be lower to that extent'.[4]

Apprentices receive on-the-job training and provide entrepreneurs with cheap supplies of labour. This encourages the adoption of labour-intensive techniques of production. However, the use of apprentice labour implies that the organisation of work in the small or informal sector firms differs from that in the large firm.

In a modern factory, tasks are usually distributed in such a way that an individual does the same job all the time. In the informal sector firms in the sample employing apprentices, tasks were not allocated this way. All the apprentices learnt about all the tasks so that eventually they came to possess the same knowledge as the master. The transmission of knowledge in this way implies that the master 'produces' his own competitors. An increase in demand is likely to lead to ex-apprentices establishing themselves in self-employment to satisfy the increased demand and compete with their own masters. Thus a continuing tendency to expand is inherent in the system.

Thus the apprenticeship system affects the technology of production and the choice of products, first through factor prices, and secondly by encouraging small-scale production units. Factor prices and scale of operation, on the other hand, determine the choice of products. Thus footwear production is appropriate to small firms employing apprentice labour. Footwear is among the products generally classified as capital-light and for this reason its production has often been recommended in developing countries. The main manufacturing operations of cutting leather, sewing, lasting and pressing require only hand-tools. Furthermore, they are simple enough to offer employment opportunities to the unskilled rural–urban migrant. In addition, these manufacturing operations do not require machinery of great capacity to achieve meaningful scale economies.[5]

5.2 Income Redistribution and the Demand for Footwear

Expenditure on footwear and the composition of demand

In section 5.1 it was postulated that an income redistribution in favour of the poor may increase employment either by increasing the total demand for footwear or by changing the composition of demand towards more labour-intensive types of footwear, or both. In this section we try to determine which of these propositions is more likely to be valid, using the data from our survey of consumer demand patterns for footwear.

Some 109 heads of households were interviewed in field work. At that time, the minimum daily wage rate was N₵4, which amounted to about N₵100 per month. Wage earners with this level of income constituted the lowest income class. However, taking into account the high rate of inflation of over 100 per cent per annum and the high prices of basic consumer goods, it is more realistic to classify all those earning less than N₵300 per month among the low-income group.

In the consumer survey, information was collected on expediture on both durable and non-durable goods. Table 5.2 shows expenditures on consumption items included in the survey. As explained earlier, the sample was based on wage earners only. This affected the data on expenditure in several ways. First, 6 per cent of expenditure on footwear is higher than the national average. In the household economic survey of 1974–5,[6] the equivalent figure for urban areas was 2 per cent. However that survey included all persons in urban areas, a group that uses footwear much less than wage earners. For clothing, the survey finding was 10.2 per cent, compared to our 9 per cent.

Another difference is that our figure for expenditure on housing is lower than the survey findings, probably because some wage earners in

TABLE 5.2   Household consumption, expenditure (per month)
of wage earners in Kumasi

	*Monthly average in cedis**	%
Food	214	57
Housing (rent)	19	5
Education	23	6
Transportation fees	15	4
Medical services	8	2
Gifts and donations	12	3
Beverages and tobacco, drinks	31	8
Clothing	33	9
Footwear	22	6
Total	377	100

* 1 cedi (N₵) = L sterling 0.16 at the time of our survey.

urban areas live in subsidised housing. The household survey also included several minor items like furniture and entertainment.

In what follows, expenditure classes are used as a proxy for income classes owing to the difficulty of estimating income from self-employment (accentuated by the political upheavals in Ghana just before our fieldwork). Table 5.3 shows the percentage distribution of the population sampled, expenditures and quantity of footwear according to the various expenditure/income classes.

For the sake of convenience, we refer to the first two classes as the low-income group, and to the middle two and the last two groups as the middle- and high-income groups respectively. Table 5.3 shows that the proportion of expenditure devoted to footwear decreases with total household expenditure, a finding consistent with Engel's law. There is no known comparable study of household expenditure structures in Ghana. In the household expenditure survey mentioned earlier, there was no information on expenditures according to expenditure or income classes. A study by Dutta-Roy[7] in 1969 which examined household expenditures according to income classes did not include footwear. In this report Dutta-Roy claimed that the proportion of expenditure on clothing increased with the level of income and thus he concluded that 'there is no level of satiety on the expenditure on clothing'.[8] Our results indicate on the contrary that as there may be a level of income beyond which consumers do not wish to spend more on footwear. We will return to this question at a later stage.

TABLE 5.3 Percentage of individuals, expenditures and physical quantity of footwear according to expenditure classes

	Monthly per capita expenditure classes*					
	100–200	200–300	300–400	400–500	500–700	700 +
Individuals in each class (%)	27.5	17.4	21.1	15.6	9.2	9.2
Cumulative (%)	27.5	44.9	66.0	81.6	90.8	100.0
Expenditure on footwear (%)	16.0	14.0	24.0	21.0	11.3	13.7
Cumulative (%)	16.0	30.0	54.0	75.0	86.3	100.0
Expenditure on footwear as % of total exp.	6.2	6.1	6.0	5.0	3.9	3.2
No. of footwear in each class (%)	22.5	16.5	20.0	15.9	11.8	13.3
Cumulative (%)	22.5	39.0	59.0	74.9	86.7	100.0

* Expenditure classes are based on total consumption expenditures, including both durables and non-durables.

When the shares of each expenditure class in population are compared with their respective shares in expenditure and the physical number of pairs of footwear possessed, a high degree of inequality is revealed. The low-income group, which accounts for about 50 per cent of the total population, is responsible for only 30 per cent of expenditures. The figures for the middle- and high-income groups are 18.4 and 25 per cent respectively. It is not therefore clear whether the high-income group possesses more footwear than the low-income group, or whether the expenditure inequality is owing to a preference for higher quality and more expensive footwear as incomes rise. A comparison of population shares and shares in the numbers of shoes purchased reveals that the latter is more equally distributed than expenditure shares. The low-income group possesses 39 per cent of all footwear compared to their share of only 30 per cent of expenditures (see Table 5.4).

Thus although the low-income group has a higher share in the physical quantity of footwear than in total expenditures on footwear, this relationship is reversed for the middle-income group. For the high-income groups, the shares are more or less equal. These data imply that as incomes rise from low to middle income, there is a dramatic shift from cheap to expensive footwear, although there is no such shift in moving from a middle to a high income.

TABLE 5.4 Expenditure on, and possession of, footwear, by income groups (%)

	Low income	Middle income	High income
Expenditure share (%)	30	45	24
Quantity of footwear (%)	39	35.9	25.1

If we assume that expensive footwear is also high-quality footwear,[9] then it appears that there is some satiety level to the expenditures which are the result of shifts to higher-quality footwear as incomes rise. The level is reached in the middle-income group. Thus the high-income group may be wearing footwear of similar quality to those in the middle-income group, but the footwear in these two income groups may differ markedly from those used in the low-income group who in general tend to buy cheap types of footwear.

The fact that there is, in general, less inequality in the distribution of footwear than in expenditures on footwear indicates that as incomes rise people increase their stock of footwear and, even more importantly, shift to more expensive footwear.

If a redistribution of income is such that it transfers incomes from the highest income group to the NȻ100–200 group, so that the latter moves into the NȻ200–300 range, then the purchase of inexpensive footwear may increase. But if the redistribution moves all those in the NȻ100–300 group into the middle group, the consumption of expensive footwear would increase at the expense of the cheap types. In both cases, therefore, the composition of demand for footwear would change. Nothing can be said, *a priori*, concerning the directions of aggregate demand for, and expenditures on, footwear. These would depend on the reaction of the high-income group which is likely to pay for the redistribution.

Thus to estimate employment effects of aggregate expenditure changes owing to income redistribution would not help the policy-maker. Even if redistribution were to increase footwear consumption in the aggregate, it would tend to do so through a change in the pattern of demand and therefore it is necessary to identify the kinds of footwear towards which demand would shift.

The demand pattern for footwear

The first problem is to find out which types of footwear are purchased by the poor, and secondly to determine whether this footwear is relatively more labour-intensive. This section concerns the first problem. (The second will be dealt with in section 5.3.)

The consumer survey attempted to establish the ownership pattern for footwear by examining the types of footwear possessed by household heads, the characteristics of the footwear that motivated their purchase, and the uses to which the footwear was put. Classifying footwear according to whether they were shoes, half-shoes or sandals, and according to the material used in their manufacture, ten footwear types were identified. The total number of footwear was 484 pairs, that is, an average of just over four pairs of footwear per head of household. In general, the average prices shown in Table 5.5 follow an expected pattern. Imported footwear, often regarded as the best quality, tends to command the highest prices. Footwear made wholly or partly of natural leather tends to be more expensive than that made from artificial leather or rubber products.

As indicated by the standard deviation, it appears that the intra-type price differences are much larger than the inter-type differences. This is because the price of footwear depends on factors other than natural or artificial leather. For instance, the method of sewing and lasting determines the durability of the footwear. Moreover, in Ghana prices may depend on whether footwear is manufactured and sold in the formal or in the informal sectors.

The large intra-type price differences imply that we cannot establish any ownership pattern for footwear in terms of the classification in Table 5.5. The cheap footwear preferred by the low-income group could be found within each type of footwear. Thus, if the low-income and high-income groups are motivated by similar considerations (characteristics and uses of footwear) when buying footwear, they may purchase the same type, with the poor preferring the cheaper ones and the rich the more expensive ones. This is shown in Table 5.6, which shows the percentage of footwear possessed by individuals in the sample, according to footwear type and according to expenditure class. Table 5.6 does not reveal any consistent ownership pattern with regard to income levels and footwear types. For instance, there is low-income group dominance in both expensive type 4 and inexpensive type 5 half-shoes. More surprising is the dominance of low-income classes of type 2 and type 9 leather shoes and sandals. Out of the ten footwear types, it is only in four types (1, 5, 7, 10) that there is a clear trend that the poor are more represented in cheaper types than the rich, and vice versa.

The fact that the low-income groups possess expensive footwear and

TABLE 5.5   Average prices* of footwear types

Footwear type	Average (mean) price per pair in cedis	Standard deviation	No. of footwear pairs
1 Imported footwear of all types	114	75	185
2 Shoes with leather upper cemented on polythene or hard rubber unit soles	73	39	58
3 Similar to type 2 but with artificial leather (PVC) uppers	54	30	12
4 Half-shoe with leather uppers on wooden sole	38	34	16
5 Similar to type 4 but with artificial leather uppers	27	11	7
6 Sandals with artificial leather uppers and wooden heel on micro-cellular rubber sole	66	50	6
7 Sandals (native) with leather uppers and leather sole	35	17	4
8 Sandals with rubber uppers and rubber sole	8	10	61
9 Sandals with leather uppers on microcellular rubber sole	37	19	51
10 Sandals with artificial leather uppers on micro-cellular rubber sole	29	12	69

* Prices are at replacement values.

indeed possess a higher proportion of some types than the high-income group implies that the classification of footwear is inadequate as a basis for distinguishing low-income from high-income footwear. In practice, it is difficult, if not impossible, to classify footwear into unique groups according to quality and prices in Ghana owing to the existence of such factors as dual distribution channels and to the fact that, for all footwear manufactured in the formal sector, there may be an imitation type in the informal sector which may be the same in outlook but having a lower price.

TABLE 5.6 % number of footwear possessed by individuals according to type of footwear and according to expenditure class

Footwear type	Expenditure class						Total
	100–200	200–300	300–400	400–500	500–700	700+	
1 Imported footwear of all types	16	14	21	19	16	14	100
2 Shoes with leather uppers cemented on polythene or hard rubber unit soles	34	17	21	10	8	10	100
3 Similar to type 2 but with artificial leather (PVC) uppers	0	0	17	33	8	42	100
4 Half-shoe with leather uppers on wooden sole	32	25	31	6	0	6	100
5 Similar to type 4 but with artificial leather uppers	72	14	14	0	0	0	100
6 Sandals with artificial leather uppers and wooden heel on micro-cellular rubber sole	25	50	0	0	25	0	100
7 Sandals (native) with leather uppers and leather sole	13	18	20	15	10	24	100
8 Sandals with rubber uppers and rubber sole	10	10	38	14	14	14	100
9 Sandals with leather uppers on microcellular rubber sole	39	10	15	16	10	10	100
10 Sandals with artificial leather uppers on micro-cellular rubber sole	26	28	14	16	12	4	100
% of household heads	27	17	21	16	9	9	100

Assuming that consumer utility is satisfied by a combination of the characteristics of footwear, we can classify the different types of footwear according to their characteristics. By comparing the characteristics of footwear preferred by the poor and the rich, we can identify the types of footwear likely to be purchased by people at different levels of income.

The characteristics of footwear that may satisfy consumer utility are that the footwear can protect the consumer's feet, it may enhance his prestige and status in society and impress others. However, these features cannot be quantified and there are no objective criteria to measure them. Some proxy measures were therefore adopted in order to derive some qualitative conclusions.

We assume that the characteristics most likely to influence consumers are durability, low price, fashion and good finish. The first two attributes are economy elements, while the last two concern prestige. In choosing footwear, an individual combines these characteristics in a way that provides the maximum utility.

Assuming that consumer choice is a function of product characteristics, it is not especially important to know the exact price of a particular pair of footwear. For our purposes, it is enough to classify as high- or low-priced. Table 5.7 shows consumer valuation of the characteristics of types of footwear identified in the household expenditure survey. The table shows that most of these scored well with regard to the prestige elements of fashion and good finishing. Taking the two

TABLE 5.7 Attribute characteristics of footwear possessed according to footwear types (%)

Footwear type	Attribute characteristics				
	Durable	Low purchase price	Fashio-nable	Good finish	Total
1 Imported footwear of all types	75	2	19	4	100
2 Shoes with leather uppers cemented on polythene or hard rubber unit soles	28	2	45	25	100
3 Similar to type 2 but with artificial leather (PVC) uppers	42	–	42	16	100
4 Half-shoe with leather uppers on wooden sole	19	13	37	31	100
5 Similar to type 4 but artificial leather uppers	–	14	29	57	100
6 Sandals with artificial leather uppers and wooden heel on microcellular rubber sole	33	33	33	–	100
7 Sandals (native) with leather uppers and leather sole	23	12	26	39	100
8 Sandals with rubber uppers and rubber sole	14	65	5	16	100
9 Sandals with leather uppers on microcellular rubber sole	30	12	32	26	100
10 Sandals with artificial leather uppers on microcellular rubber sole	10	23	33	34	100

characteristics together, seven out of the ten footwear types scored more than 50 per cent.

Only type 8 is rated high with regard to low purchase price. Imported footwear, on the other hand, is judged to be outstanding with regard to durability; it is followed by types 3, 6 and 9, in that order. Three out of the ten are rated sufficiently high with regard to the economy attributes of durability and low purchase price. These are types 6 and 8.

For the low-income group who can only afford a smaller quantity of footwear than the rich, the problem is to find footwear that is cheap and also combines several of the above characteristics. For the high-income group, this may not be an acute problem since it is in a better position to purchase a greater quantity of footwear, each having differing characteristics.

In this regard it is important to note that with the exception of footwear type 6 (and also to some extent type 8) there are no types of footwear that combine durability and low purchase price in a balanced way. Most of the shoes rated high for durability score badly on price, whereas type 8 which is reported to have a very low purchase price, is regarded as non-durable to some extent. Thus an important first conclusion is that there may be very few types of footwear 'appropriate' to the satisfaction of the needs of the low-income groups.

Table 5.7 also enables a comparison of the durability scores in relation to low purchase price on the one hand, and fashionable/good finish with low purchase price on the other. The footwear types that score high on fashionable/good finish tend to score better on low purchase price than the scores of durable footwear on low purchase price. This implies that there are more types of footwear that are both fashionable with good finishing and these are also relatively cheap than those that are both durable and cheap. It is apparently more difficult to produce footwear that is both durable and cheap. It may also be that such footwear is not preferred if it is not considered to be sufficiently fashionable or well-finished.

Which of the attribute characteristics influence footwear choice in the various expenditure classes?

On the basis of the footwear that an individual had, he or she was requested to specify the most important characteristic of footwear purchased. The answers are recorded in Table 5.8 which shows the attributes of footwear owned by the heads of households in the sample according to expenditure classes. The table shows that in the population as a whole, 41.6 per cent of footwear was purchased because it was thought to be durable and 10.8 per cent because of its low purchase

TABLE 5.8   Attributes of footwear owned according to expenditures class (%)

Attributes	100– 200	200– 300	300– 400	400– 500	500– 700	700+	Total
Durable	32.7	36.2	39.8	45.7	50	53.1	41.6
Low purchase price	3.7	8.7	19.4	14.5	11.1	7.8	10.8
Fashionable	37.4	32.5	25.5	27.1	16.7	15.6	27.4
Good finish	26.2	22.5	15.3	12.9	22.2	23.4	20.2
Others	–	–	–	–	–	–	–

Column elements may not add to 100 owing to rounding.

price. Fashion and good finish (prestige) accounted for the purchase of 47.6 per cent of total footwear. This implies that prestige, status and good impression – which in our study are represented by fashion and good finish – are important reasons for footwear purchase, followed closely by durability. Low purchase price does not seem to be significant. Taken together, however, the economic attributes are the most important, accounting for 52.4 per cent of all purchases.

Since our aim is to determine the income effects on the choice of footwear characteristics, it is important to examine the distribution of attributes according to expenditure classes. It can be seen that durability is more important at the higher levels of income; conversely, fashion and good finish are less important. Similarly, low purchase price is also less important to the low-income consumers than to the high-income group. This implies that the prestige characteristics are equally important, if not more important, to the poor in their selection of footwear types. The economic attributes together amount to only 36.4 per cent, while 'fashionable' and 'good finish' account for 63.6 per cent.

These results are unexpected. One would have expected that the poor would appreciate durability and low purchase price more than the rich. Prestige and status-enhancing are characteristics that could be regarded as redundant, whereas durability and low purchase price are characteristics that are essential to 'appropriate' footwear needed to satisfy the needs of the poor. Thus, surprisingly, the poor seem to prefer 'redundant' characteristics even more than the rich. The finding throws into doubt the wisdom of attempting to develop 'appropriate' footwear products (or any similar product) by removing excess or redundant characteristics from existing ones to satisfy the needs of the poor. Therefore, we need to examine the results more closely. In particular the seeming insignificance of low purchase price as a motivation in footwear purchases among the low-income group requires explanation.

It appears that price is not a crucial factor in consumer decisions of those in low-income groups: other factors such as good finish and fashion are more important. Another possible explanation is that when the price of particular footwear is considered in relation to its quality, the price may not appear to the purchaser to be low. This is particularly important because many pairs of footwear possessed by the low-income group are considered by their owners to be of low quality.

An explanation of the fact that economy attributes, as a whole, are less important to the low-income groups may be derived from our earlier analysis.

It appears that 'appropriate' footwear may not exist. As shown earlier, durable footwear tends to be very expensive. This is not the whole story, however, since there are at least two types (6, 8) that are rated high in the economy attributes of durability and low purchase price taken together. Therefore consumer preferences may also be a contributory factor.

An explanation for this phenomenon may lie in the uses to which footwear is put. Table 5.9 shows the uses to which the ten types of footwear are put. All footwear types are to a varying extent used for work and for casual wear. Only four (types 1, 2, 3, 7), however, are judged to be suitable for ceremonial occasions. (Ceremonial occasions include funerals, outdoor ceremonies, church-going, etc. Such ceremonies are extremely important in the social structure of Kumasi.) Native sandals (type 7) are predominantly used for traditional ceremonies like funerals, durbars, etc. Only types 2, 3 and 6 are put to all four uses to a significant degree.

Among the footwear types rated high in terms of economy attributes, only type 6 appears to be suitable for ceremonies, work and for casual wear. Type 8 is predominantly used for casual wear.

Footwear type 6 is therefore the only all-purpose footwear which is both cheap and durable. From the point of view of future product development in the industry, type 6 is of particular interest. It also has the advantage that it is one of the few footwear types that uses local raw material – the heel is made of wood produced in Ghana.

The next question concerns the uses to which the various income classes put their footwear. Table 5.10 shows the uses to which footwear in the sample were put to according to level of expenditure/income. On the whole, ceremonial use was the single most important reason why footwear was purchased (last column), accounting for 42.4 per cent of all uses. Table 5.10 also reveals that ceremonial use was relatively more important in the low-income group of NȻ300 and below than in the

TABLE 5.9　Usage of footwear according to type of footwear (%)

Footwear type	For ceremonial occasions	Work	Casual wear	Total
1 Imported footwear of all types	64	27	9	100
2 Shoes with leather uppers cemented on polythene or hard rubber unit soles	33	44	23	100
3 Similar to type 2 but with artificial leather (PVC) uppers	42	25	33	100
4 Half-shoe with leather uppers on wooden sole	–	69	31	100
5 Similar to type 4 but artificial leather uppers	–	57	43	100
6 Sandals with artificial leather uppers and wooden heel on microcellular rubber sole	33	33	33	100
7 Sandals (native) with leather uppers and leather sole	78	12	10	100
8 Sandals with rubber uppers and rubber sole	5	30	65	100
9 Sandals with leather uppers on micro-cellular rubber sole	14	40	46	100
10 Sandals with artificial leather uppers on micro-cellular rubber sole	6	38	56	100

high-income group of N₵500 and above. Since footwear for ceremonial occasions needs to be relatively fashionable and have good finishing, it seems that the explanation of its importance in determining consumer demand may be found in the uses to which footwear is put. Since a higher proportion of footwear owned by the low-income group is used for ceremonial occasions, a higher proportion is also purchased for reasons of fashion and good finishing.

Thus if a poor person in Kumasi buys only one type of footwear, he may choose a type suitable for ceremonial uses.

In terms of the Lancaster demand model, the preference of the low-income group for prestige rather than for economy factors could perhaps be explained in terms of inefficient choices, which fail to

Table 5.10   Use of footwear according to expenditure class (%)

| | Expenditure classes (%) | | | | | | |
	100–200	200–300	300–400	400–500	500–700	700+	Total
For ceremonial Occasions	43.0	47.7	41.8	39.2	35.2	46.8	42.4
Work	34.0	27.0	30.6	34.8	40.7	29.8	32.1
Casual wear	23.0	25.3	27.6	26.0	24.1	23.4	25.5

combine attribute characteristics that give maximum utility. However, such an explanation is not plausible in the light of our finding that the characteristics of the footwear purchased are consistent with the uses to which it is put.

A difficulty facing consumers is that no type of footwear combines all the characteristics required. Footwear that is durable and elegant tends to be so expensive that the low-income group cannot afford it. On the other hand, fashionable/good finish footwear is relatively cheaper than durable footwear.

When the phenomenon is explained in this way, the argument that it is possible to develop an appropriate product to satisfy the needs of the poor seems reasonable. The trouble is that when appropriate footwear is defined to include both economy and prestige characteristics, it becomes meaningless. The definition of 'appropriate' footwear in a society like that in Kumasi may be more complex than is usually assumed. If appropriate footwear is defined only in terms of durability and low price, there may be no market for it. If other attributes are added, it may be difficult to produce such footwear at sufficiently low prices.

The present purchasing behaviour of low-income groups is influenced by the availability of different types of footwear, and even more by consumer preferences, with the latter in turn being determined by the uses for footwear.

The footwear favoured by an income redistribution towards the poor is likely to be that which, given its low price, also has the attributes of elegance and durability.

Redistribution effects on demand for footwear

Because footwear is not a homogeneous commodity and because prices reflect not only quality but also the channels of distribution, the calculation of Engel elasticities based on the expenditures on footwear did not serve any useful purpose. Several consumption functions were estimated but none fitted the data satisfactorily. The function that fitted the data best was the log-reciprocal[10] which corresponds to the hypothesis that income elasticity is inversely proportional to income. However, the R^2 was only 0.23 and the values of Student's $-t$ were low. These estimates are not therefore acceptable as a basis for the calculation of Engel elasticities.

A better method might have been to estimate functions for each type of footwear in as homogeneous a manner as possible, but the data base did not permit this. Alternatively, one could estimate consumption functions for each expenditure class, assuming that intra-class variations in footwear quality and prices as a result of distribution channels are minimal. However this was impossible because there were too few observations in each class. It was therefore decided to use fairly simple methods to furnish broad estimates. The calculations are based on the physical quantities of footwear possessed by each expenditure or income class.

Since individuals in a given expenditure class are likely to use the same distribution channel (formal/informal) and to buy footwear of a similar quality, it can be assumed that the quantity of footwear an individual possesses is determined by the expenditure class to which he or she

TABLE 5.11 Transfer of 3.9% of aggregate expenditures from the highest expenditure class 6 to lowest expenditure class 1

| | *Expenditure class** (N¢ per month)* | | | | | | |
	1 *100–* *200*	*2* *200–* *300*	*3* *300* *400*	*4* *400–* *500*	*5* *500–* *700*	*6* *700+*	*Total*
Before transfer							
Per capita expenditure	167	221	340	446	629	1 050	
Total footwear	109	80	97	77	57	64	484
No. of footwear							
per person	3.6	4.2	4.2	4.5	5.7	6.4	
After redistribution							
Per capita expenditure	221	221	340	446	629	888	
Total footwear	126	80	97	77	57	64	501
No. of footwear							
per person	4.2	4.2	4.2	4.5	5.7	6.4	
% increase in total footwear							3.5
% increase in class 1 footwear							15.6

* Expenditure classes are based on total consumption expenditures, including both durables and non-durables.

belongs. We have also assumed that the aim of redistribution is to upgrade incomes in the lowest income group 1 (NȻ100–200) to the levels in group 2 (NȻ200–300), and that all the transfer will be taken from the highest income group 6 (NȻ700+). The transfer needed to accomplish this amounts to 15 per cent of the expenditures of the highest expenditure group but only 3.9 per cent of aggregate consumption expenditure.

The results of the exercise are shown in Table 5.11. If income levels in class 1 are upgraded to the level in class 2, the quantity of footwear possessed in class 1 would be increased by 15.6 per cent, or by 3.5 per cent of the total number of footwear possessed by all classes. In other words, a transfer of 3.9 per cent of aggregate expenditure to the lowest income group would increase demand for footwear by nearly as much. The footwear possessed by class 6, which bears the whole burden of the transfer, would not be affected because the income transferred would not be enough to move them into a lower expenditure class.

5.3 Income Redistribution and Employment Creation

In the preceding section it was shown that the low expenditure classes 1 and 2 earning not more than NȻ300 per month tend to buy cheaper footwear than the middle- and high-expenditure classes. It was also estimated that a transfer of 3.9 per cent of aggregate expenditure to the lowest income group could lead to increases in footwear demand of 3.5 per cent in all expenditure classes, and 15.6 per cent in expenditure class 1. Since the increase in demand would affect only the cheaper kinds of footwear, it is necessary to examine which enterprises produce such footwear, bearing in mind that firms of different sizes employ different production techniques, and varying numbers of people.

The seventy-one enterprises in our sample differed markedly with regard to the amount of capital equipment, methods of production and types of labour employed. The value of capital equipment per enterprise ranged from NȻ40 to over NȻ300 per enterprise. Most of the small firms had no electricity or water. However, the footwear produced by small and large firms is quite similar.[11] This may be because the small firms are also adopting mass-production methods, as now described.

To study manufacturing technology in the footwear industry, it is convenient to separate the manufacturing processes into the main stages of cutting, sewing, cementing, lasting, pressing and finishing. All these processes can be accomplished by machine or manual operations or by a combination of the two. Cutting leather uppers, for instance, may be done using hand-tools costing no more than N₵10 or power presses worth over N₵30 000.

One process that is highly mechanised is sewing. Even among the small informal firms in which manual operations are commonplace, sewing tends to be accomplished by machines in small informal firms which have specialised only in sewing uppers for other footwear producers. Producers have found that it is cheaper to have their uppers sewn on by the specialised firms. Unit soles made of wood are also cut by machines in special enterprises. Even the pressing stage is becoming rapidly specialised. Thus in the small-scale sector, more than one firm is often engaged in the production of a particular type of footwear with each firm having only a small fixed capital requirement.

The stages of production that are becoming specialised are those where the advantages of mass production are most obvious. Cutting and shaping unit soles, sewing and pressing are typical examples. However, cutting uppers may be done individually since designs differ. In the sample, 85 per cent applied manual operations in cutting; the others used machines. The equivalent manual operations for the other stages are sewing (4 per cent), cementing (90 per cent), lasting (90 per cent), pressing (64 per cent).

Thus although many firms may apply manual operations, only a few types of footwear are produced by manual systems alone in Kumasi. The two types of operations are combined in the production of most types of footwear. This in turn implies that footwear, particularly sandals, produced in the labour- and capital-intensive sectors may not differ much with regard to quality.

Footwear purchase pattern and size of firms

Although several studies[12] have concluded that the cheap footwear which low-income groups can afford could be quite profitably produced on a large scale, casual impression tends to support the view that low-income classes buy footwear produced by small informal-sector enterprises, while the rich buy products from large formal enterprises. We attempted to establish the validity of this impression by collecting ex-factory and retail price information from small and large enterprises producing similar types of footwear.

The types of footwear produced by both large and small enterprises were types[13] 6, 9 and 10 (described in Table 5.5 above). These three types of footwear accounted for over 75 per cent of footwear production by all enterprises in the sample. The small enterprises had machinery/tools and furniture valued at less than N₵3000 per enterprise. The equivalent value of machinery/tools and furniture in the large enterprises ranged from N₵12 000 to over N₵300 000.

An examination of data on capital–labour ratios, and ex-factory and retail prices, for all the enterprises in the sample revealed that the capital–labour ratio is invalid as a predictor of either kind of price. Ex-factory footwear prices were found to be similar – N₵32 in small firms and N₵31 in large ones. On the other hand there seemed to be a significant difference in average retail prices – N₵34 and N₵52 in small and large firms respectively. Since the products of small enterprises are retailed at lower prices than those of large ones, and since low-income groups buy cheap types of footwear, it is plausible to conclude that the poor would largely buy footwear made by small enterprises.

It appears that although small enterprises employ unpaid apprentice labour, they do not enjoy any cost advantage over large enterprises. The more advanced technology applied in large enterprises overcompensates for their higher wage bill. This finding confirms the studies of McBain and others.[14]

However, the studies showing that cheap footwear could be most profitably produced in large plants often fail to include distribution and marketing costs in their cost estimates. A comparison of Tables 5.9 and 5.10 indicates that the price advantage of the small firms is the result of lower marketing costs, because such enterprises tend to sell directly to the consumer. Thus ex-factory prices may not differ greatly from retail prices.

Secondly, small firms also tend to have low overhead costs. On the other hand, the products of large firms are often sold in stores with high overhead costs. Where such products are retailed by small shops, they frequently originate from relatively large wholesalers who, in turn, supply the retailers. Thus the products of large firms may pass through many hands before reaching the consumer. It is therefore necessary in simulation studies to examine not only the product's costs but also the costs at which the product would reach the consumer – the ultimate objective of production. The differential market structure in developing countries makes this essential.

In section 5.2 it was estimated that a transfer of 3.9 per cent aggregate consumption expenditure to the lowest expenditure class, accounting for 27.5 per cent of the sample, could mean that the number of pairs of footwear per capita in that income class would change from 3.6 to 4.2 per capita. In other words there would be an additional demand for footwear of 0.6 pairs per person (see Table 5.11).

The population of Ghana is estimated at 10 million, of which some 5.4 million are of working age and, according to our survey, some 1.5 million (27.5 per cent) are in the low-income group earning less than N₵200 per month. If all the 1.5 million persons increased their demand for footwear by 0.6 per capita, an additional 900 000 pairs would be needed. If the increased demand of their dependants is added, then 1–1.5 million pairs would be needed.

The above estimates assume that the survey findings are valid for both rural and urban areas, although our survey was in fact based on an urban centre and on wage earners only. Incomes are lower in rural areas and in each income class footwear per capita may be low in comparison with urban areas. A household survey in 1974–5[15] revealed that although urban areas accounted for only 30 per cent of the total population, they also spent about 70 per cent of total expenditures on clothing and footwear.[16] The report also showed that 1.5 per cent of income was spent on footwear in rural areas, as compared with 1.9 in urban areas.

More importantly, our sample was based on the section of the population that uses footwear most. This implies that our demand projections may be higher than would be the case if non-wage earners and rural people were included in the sample. The lack of data has made it impossible to adjust for this.

Employment effects

Since the demand for the products of small enterprises is more likely to increase in consequence of a programme of income redistribution, the employment effects of such a programme may be examined largely in relation to the techniques of production applied in small firms alone, with due consideration to the indirect effects on production in large firms. The footwear industry in Ghana contains a diversity of technologies ranging from the single-operated shoemaker workshop to the most modern factories.

Table 5.12 shows the characteristics of production technologies applied in small and large footwear manufacturing enterprises in Kumasi. The small firms are both labour- and apprentice-intensive, implying that a unit increase in output generates increased employment, the greater percentage of which would be apprentice labour.

The proportion of apprentices in small and in large firms indicates the labour cost advantage of the small enterprises. Apprentices who are not paid regular wages account for 66 per cent of employment in small

TABLE 5.12 Characteristics of production techniques applied in small and in medium/large footwear manufacturing enterprises in Kumasi

Characteristics	Small	Medium/large
Machinery/tools and furniture per enterprise (N¢)	500	37 000
Persons engaged per enterprise (total)	3	12
Apprentices per enterprise	2	1.5
Capital–labour ratio	200	3 000
Annual output of footwear pairs per enterprise	4 000	50 000
Annual output of footwear pairs per person engaged	1 300	4 200

enterprises. Instead, apprentices have to pay both entrance and exit fees in return for training and daily 'chop money'. As a result of the use of apprentice labour, the small enterprises apply labour-intensive techniques of production, as shown by the capital–labour ratios.[17]

If we assume that the demand for the additional 1–1.5 million pairs of footwear is satisfied within one year, and also assuming constant output–labour ratios and that all the extra footwear is produced in small enterprises, then 300–400 small establishments would be required employing a total number of 800–1200 persons.

Since the footwear and leather industry in Ghana employs some 12 000 persons (about 2.4 per cent of total manufacturing employment), this increase in demand could increase employment in footwear production by nearly 10 per cent and total manufacturing employment by 0.24 per cent. Some 66 per cent of the employment created in small industry would be for apprentice labour. The increased demand for footwear could also be satisfied by between twenty and thirty large firms employing 300–400 persons, with a much lower effect on direct employment. However, such firms may generate more indirect employment through the distribution system.

The fact that the increased demand justifies the establishment of between twenty and thirty large firms instead of 300–400 small ones is important, because if the large firms can produce at lower prices because of scale effects, income redistribution of this sort will not benefit the small firms.

However, our study indicates that as long as small firms have lower marketing costs, they may be able to sell at lower retail prices than large ones. In other words, the scale disadvantage of small firms may be offset by nearness to the market.

Considering that the small-scale sector accounts for about 80 per cent of total employment in the footwear industry, and furthermore that only 3.9 per cent of aggregate expenditures was transferred, the aggregate employment increase is not insignificant. However, increasing the amount of expenditures transferred much above the 3.9 per cent level may not increase employment significantly. This is because as income increases from low to middle levels there is a marked shift towards more expensive types of footwear. A large transfer that moves the low-income groups into the middle-income groups may lead to a considerable increase in expenditures on footwear but not to a significant increase in the quantity of footwear purchased. Since there is no evidence that expensive footwear is more labour-intensive than cheap footwear – our study shows that the contrary is more likely to be true – such a transfer may not increase employment much more than our earlier estimates indicated.

Employment effects on unskilled labour

It was postulated in section 5.1 that those in the low-income bracket may be purchasing footwear produced by small firms and by unskilled labour; that small firms are more likely to benefit from any income redistribution programme has already been discussed. The part of the hypothesis concerning unskilled labour remains to be investigated.

The question is whether or not the small firms that produce footwear for low-income groups also offer the greatest employment opportunities for unskilled labour. It has sometimes been argued, by Albert Hirschman for example, that labour-intensive methods of production with their operator-paced processes have a greater need for skilled labour than capital-intensive methods in which the process is machine-paced. However the reality in the small and large firms examined in this study seems to be the opposite.[18] In both the small and in the large firms, workers become skilled after they have been engaged on a particular process for a period of time. Thus we need to examine whether workers in the small-scale sector at the beginning of their employment are more unskilled than those in the large-scale sector also at the beginning of their employment.

It was argued in section 5.1 that to become self-employed, one must have a complete knowledge of the trade. The master craftsman is such a person. He is likely to possess more of the skills needed in footwear production than the potential worker in the large firms. However, only one person with complete knowledge is needed in the typical small informal firm. As we have seen, most of the labour force are apprentices, having little or no knowledge of the trade. The majority are young with low levels of education. Because of this, the relationship between master and apprentice is structured around the transmission of knowledge rather than remuneration, and productivity is necessarily less important than in wage employment.

The low demands that the apprenticeship system makes on educational standards and previous experience imply that unskilled labourers can easily become apprentices. Indeed, this is the entry point to the urban economy for many migrants from rural areas.

The minimal demands of the apprenticeship system may be contrasted with the qualifications required for employment in large firms. In Ghana, a minimum level of education to middle-school level, that is, about ten years of formal education, is required. Many potential employees have received some kind of technical or vocational training and they are normally in their prime working age. Thus, through the apprenticeship system the small-scale sector, although labour-intensive, offers more employment opportunities to unskilled labour than the large-scale sector.

Since low-income groups tend to purchase more of their footwear from the small-scale sector, it follows that their purchase pattern is skewed towards footwear produced by apprentice labour rather than wage labour. Thus the consumption pattern of the low-income group may not be invariant as between skilled and unskilled labour according to our definition, and income redistribution to the poor may encourage the employment of unskilled labour.

5.4 Summary and Conclusions

In this chapter we have shown that because of the uses to which footwear is put in Ghana both rich and poor have similar purchase patterns. The main difference is that the rich tend to buy more expensive shoes whereas low-income groups make do with cheaper ones.

The low-income groups have a strong preference for prestige-enhancing attributes in footwear. This leads them to prefer footwear that is fashionable rather than that which is only durable. This finding may pose a problem for the supporters of appropriate product strategies. Low-income groups tend to buy from small-scale firms, which have lower marketing costs, charge lower retail prices and apply relatively labour-intensive techniques. However, large firms create more indirect employment through the distribution system.

The study underscores the importance of disaggregating production units according to scale because the purchase pattern of the poor may not be invariant to the sizes of manufacturing firms, and firm sizes are associated with particular types of techniques of production.

Furthermore, since the small-scale sector mainly employs apprentice labour and since unskilled labour has a better chance of entering apprenticeship than wage employment in the large firms, an income redistribution would tend to increase the employment of unskilled labour through the apprenticeship system.

Due to the change to higher-quality footwear as incomes rise, rather than the consumption of greater quantities of a similar type of footwear, income redistribution effects on employment in the footwear industry may be high only when the beneficiaries are the poorest sections of society.

Notes

1. The author was a faculty member, University of Ghana, when this study was prepared; he is now with the ILO's Southern African Team for Employment Promotion, Lusaka, Zambia.
2. See George Aryee, *Small-scale Manufacturing Activities: A Study of the Inter-relationships Between the Formal and the Informal Sectors in Kumasi* (Geneva: ILO, 1977; mimeographed World Employment Programme research working paper; restricted).
3. Ibid.
4. A. K. Sen, *Employment Technology and Development* (Oxford: Clarendon Press, 1975) p. 37.
5. See E. Staley and R. Morse, *Modern Small Industry for Developing Countries* (New York: McGraw-Hill, 1965) Table 7.1.
6. *Summary Report on Household Economic Survey* (Accra: CBS, 1979).
7. See D. K. Dutta-Roy, *The Eastern Region Household Budget Survey* (Accra: ISSER Legon, 1969).
8. Ibid, p. 81.
9. This needs to be qualified. The high-income group may be paying high prices for their footwear not only because of high quality but also because they earn more than the low-income classes and buy their footwear in the formal sector's distribution channels with high overhead costs.
10. That the log-reciprocal model provided the best fit was expected, since this model implies a saturation level. As has been shown earlier, the demand for footwear measured by expenditures approaches a level of satiety somewhere between the middle- and high-income classes.
11. This is particularly true of sandals in Ghana. McBain also found that a wide range of labour or capital-intensive technologies can be used to produce a given footwear type. See N. McBain, 'Developing Country Product Choice:

Footwear in Ethiopia', *World Development*, vol. 5, nos. 9–10 (1977). The findings refute the claim by Frances Stewart that once a product is specified, the choice of technology is eliminated. See F. Stewart, 'Choice of Technique in Developing Countries', *Journal of Dev. Studies*, vol. 9 (Oct 1972).

12. See, for instance, N. S. McBain and J. Pickett, 'Footwear Production in Ethiopia – A Case Study of Appropriate Technology', *Journal of Modern African Studies*, vol. 13, no. 3 (1975). Similarly, Marsden claimed that the establishment of a plastic footwear factory in a developing country made 5000 small-scale shoemakers unemployed: see K. Marsden, 'Progressive Technologies for Developing Countries', *International Labour Review*, vol. 101 (May 1970).

13. Type 6: sandals with artificial leather uppers on wooden heels with microcellular rubber sole; type 9: sandals with leather uppers on microcellular rubber sole; type 10: sandals with artificial leather uppers on microcellular rubber sole.

14. McBain and Pickett, 'Footwear Production in Ethiopia – A Case Study of Appropriate Technology'.

15. See *Summary Report on Household Economic Survey, 1974–75* (Accra: CBS, 1979).

16. Ibid, Table 16.8.

17. There is evidence that in the small-scale, informal footwear and in other manufacturing sectors, apprentice labour is employed as a substitute for capital. See George A. Aryee, 'Labour Utilisation and Factor Productivity: Evidence from Non-wage Activities in Kumasi, Ghana', in G. Mikkelsen (ed.), *Small-scale Industry in Africa* (Uppsala, Sweden: Scandinavian Institute of African Studies, forthcoming).

18. The studies of the footwear industry in Ghana and in Ethiopia by McBain also showed that capital-intensive products tend to be associated with a relatively high share of skilled labour in total labour requirements; see ILO, *The Employment Implications of Technological Choice and of Changes in International Trade in the Leather and Footwear Industry*, Technical Report III for Second Tripartite Technical Meeting for the Leather and Footwear Industry (Geneva: ILO, 1979).

6 Product Choice and Employment in Furniture-making in Kenya

William J. House[1]

6.1 Introduction

The objective of this chapter is to test the income distribution–technology–employment hypothesis for furniture in Kenya, by exploring the link between the suppliers of the commodity and the income characteristics of the purchasers. The choice of furniture is especially appropriate, first in the light of the emphasis placed in Kenya's Development Plan on the goal of satisfying the basic needs of the poor[2] and secondly because furniture is specifically mentioned as a basic-needs good in the Declaration of the ILO's World Employment Conference which strongly endorsed the principle of seeking to satisfy the basic minimum requirements of the poor.

The basic hypotheses can be restated in terms of the furniture industry in Kenya. The existing distribution of income determines a pattern of expenditure whereby low-income consumers purchase items of furniture to satisfy basic needs, and these goods are produced in a relatively labour-intensive manner. In contrast, higher-income consumers purchase items of furniture of a generally superior quality, embodying characteristics superfluous to the satisfaction of basic needs. These products are much less labour-intensive than the furniture consumed by the poorer groups. Consequently, a redistribution of income would increase the demand for the lower-quality items and thus cause an increase in total employment in the low-wage sector of the industry. The secondary expenditure effects will have induced a further increase in incomes which strengthen the initial income redistribution.

The basic-needs characteristics of items of furniture are fairly obvious. Items such as chairs, sofas and beds must provide comfort for

148

the body of those wishing to relax or sleep. Three factors are involved in providing comfort: the size and relative proportions of the chair or bed, the shapes of its component parts and the degree of softness of the upholstery.[3] In addition, tables, stools and cupboards should be capable of providing a basic service as receptacles of food, clothing and household items so that a minimum level of hygiene and convenience can be attained. In all cases, furniture should be durable and be able to withstand multi-purpose use by poorer households.

It is impossible to quantify a minimum index of basic-needs satisfaction for furniture as can be done for food products, which may or may not fulfil minimum dietary requirements for a given level of expenditure. Therefore it is assumed in this study that all furniture sold in Kenya will incorporate the minimum necessary quality to meet these basic requirements.

However, very many non-basic characteristics will be embodied in furniture products. The quality, texture and finish of the wood, and the design, padding and upholstery used, all relate to something elusive and intangible which may be termed 'quality'. In some cases these characteristics will satisfy a degree of comfort and convenience over and above the minimum required while, in others, aesthetic and perhaps snob values will be catered for.

If poor consumers in developing countries act rationally such characteristics are luxuries, with an income elasticity of demand exceeding one. Therefore, the lowest-quality and lowest-price furniture items will be appropriate for low-income households since they are most likely to embody only those essential characteristics that are relevant to satisfying basic needs.

6.2 The Furniture Industry in Kenya

The production of furniture in Kenya is widespread. In the rural areas, where 85 per cent of the population resides, the small-scale, self-employed and often part-time carpenter is ubiquitous, constructing basic items of furniture for home use and for the market. While there are no reliable estimates of the numbers employed in the rural industry, a recent survey revealed that at least one-half of the sampled households were engaged in one or more non-farm activity and, of these, 1.5 per cent were engaged in furniture.[4] The estimated 12.5 million rural Kenyans make up 1.8 million families, since average rural household size is estimated as seven persons. This implies that at least 27 000 rural families are involved in some way in the furniture industry.[5]

In 1977 slightly more than 4000 persons were employed in the formal, modern[6] or enumerated sector while urban informal-sector employment exceeded 2000 in 1975. Earlier work by this author has shown that the Kenyan furniture industry is very competitive.[7]

Firms in the industry appear to fall into one of four different categories, depending on the quality of the product and the firm size. The smallest firms, producing the lowest quality of furniture, are found in the rural areas and in the urban informal sector.[8] Formal-sector firms with up to fifty workers are involved in making low- to medium-quality products which are competitive with informal-sector output at the low end of the quality scale. On the other hand, their better-quality products compete with those of the few large-scale factories, which employ more than 100 workers and are engaged in the mass production of relatively standardised products. The fourth category of firms appears in the 50–99 employees size-class; they are mainly engaged in producing expensive 'antique' reproductions or quality 'modern' items and might well be called the custom or craft sector.[9]

The formal workshop sector, housed in permanent buildings and employing up to fifty workers, represents a higher scale of technology than that of the rural and urban informal sectors. The main differences arise from the electrically driven machines used by this sector and the fact that their operators occupy permanent buildings.

Despite their low quality, the products of the rural and urban informal sectors appear as appropriate for serving the basic needs of poor Kenyan families. Yet, the evidence suggests that such needs are still not fulfilled for a large part of the population. Livingstone[10] painted a bleak picture of the quantity and quality of household possessions in both Tanzania and Kenya, and his findings have been confirmed by the results of the Integrated Rural Survey in Kenya. For example, while the average household has seven members, its average number of beds is only 2.71, of chairs only 3.73, and of wardrobes 0.06.

While the number of items purchased rises with income, the major constraint on the expansion of that part of the industry producing simple furniture appears to be the low incomes of the majority of the population.

Following Stewart[11] we shall examine various aspects of the technology employed in the different parts of the industry, including scale, material inputs, labour inputs and wages, and investment requirements. The nature and quality of the different products dictates the kinds of technologies employed, but the whole industry is very labour-intensive compared with other manufacturing industries.[12] Table 6.1 is based on

TABLE 6.1 Indicators of technology employed by size-class of firms

	Informal sector	1–4	5–19	20–49	50–99	100 +	Total formal
1 No. of firms	NA	44	94	22	10	6	176
2 No. engaged	NA	392	957	718	682	973	3 722
3 Value added (£'000)	NA	219	930	757	975	872	3 754
4 Value added as % total value added	NA	6	25	20	26	23	100
5 Value added per worker (K£)	362	559	971	1 056	1 431	896	1 008
6 Average wages (K£)	166	528	601	649	684	598	622
7 Material inputs (K£) per worker	NA	1 565	1 639	1 938	2 606	1 308	1 780
8 Fuel and electricity per worker (K£)	NA	NA	15.4	9.9	19.1	15.8	15.1
9 Capital: output	0.7						
10 Capital: labour	133						
11 Income of business head (K£)	707						

NA Not available.

Source. Census of Industrial Production, 1972 (Central Bureau of Statistics, Nairobi, Kenya, 1978); and W. J. House, 'Survey of Nairobi Informal Sector 1977' (Institute of Development Studies, Nairobi, 1978).

the most recent Census of Industrial Production (for 1972) and all values have been inflated to 1977 prices. Stewart has argued against the use of market prices to value different products, because of the presence of advertising, monopoly power, undesirable income distribution, etc. However, because of the presence of such a heterogeneous collection of products in Kenya's furniture industry, no other alternative was available. In any case, as was argued earlier, some of these possible distortions, such as advertising and monopoly power, are at a minimum in the industry.[13]

Following work by Nelson[14] on Colombia, it has been suggested that a number of techniques of production are in operation in any industry, so that different firms are on different production functions at any one time. The sub-group using later vintages is identified as 'modern' and these firms are likely to be large, for two reasons. Economies of scale will be more important for modern technology and the advantages of such technology are likely to be greater at relatively high capital–labour ratios. The other group of firms is identified as 'craft', using less modern equipment, employing labour of low skills and creating far lower value added per worker.[15]

Table 6.1 shows that such stylised facts are not borne out in the case of furniture in Kenya. At the lowest level of labour productivity in row (5) are found the informal-sector and tiny formal-sector workshops in the 1–4 size-class. But labour productivity in the very largest firms is less than the average for the 5–19 and 20–49 size-classes. Productivity appears significantly higher for the ten firms in the 50–99 size-class, the products of which have been identified to be of a very different quality from the rest of the industry. And these firms produce over one-quarter of the total value of output of the industry.

Average wage levels follow the pattern of labour productivity across size-classes, with the highest wages paid to workers in firms in the 50–99 size-class, presumably reflecting skill differentials. These firms also show the greatest value of material inputs per worker, reflecting the quality of the materials necessary for the characteristic finish of the end products.

While no capital stock data are available for the formal sector, estimates are made of the value of fuel and electricity used per worker, which should reflect relative differences in capital per worker across size classes of firms. The average for all formal-sector firms is a meagre K£15.1 compared with an average for all manufacturing of K£126.4, again confirming the relative labour intensity of the furniture industry. The impression gained from this exercise is that the firms in the 50–99 class use more power equipment per worker than any of the other firms. Interestingly, the very largest firms appear no more capital-intensive than those firms in the 5–19 class by this measure.

The data derived from the informal-sector survey show that labour productivity is only a third of the average for formal-sector firms, while average wages are significantly lower. Fuel and electricity used per worker are negligible. These data reflect the lower-quality output and skills employed in the informal sector. Yet the net average annual earnings of the business heads of K£707 compare very favourably with what these same persons would earn as employees in wage employment in the formal sector.[16]

While no aggregate data are available for formal-sector furniture, estimates for the whole of the formal sector suggest, for the capital-output ratio, a range from 2.5 to 1.1, and for the capital–labour ratio a value of K£2000.[17] Our estimates for informal-sector furniture are much lower for both ratios.[18]

Outside of the 100 + size-class of firms the amount of machinery used is minimal. In the largest firms there is a degree of mass production of standardised components as well as the manufacture of metal parts for what is essentially wooden furniture. The quantity and diversity of equipment grow in accordance with the size-class of firms.

In general it is difficult to imagine the industry becoming more labour-intensive, since the quantity and quality of the machinery is dictated by the design and quality of product choice. The sawing, smoothing and sanding machines in the larger firms are essential to maintain production runs of 'middle'-quality furniture. Yet operations such as glueing, carving and packaging are all carried out by hand.

6.3 Consumer Purchasing Patterns and Income Distribution

Evidence on income distribution in Kenya is very scarce. Using a study carried out in 1969 by a researcher from the World Bank, Chenery and his associates classified the degree of inequality as 'high'.[19]

For purposes of constructing its Nairobi cost of living indices the Central Bureau of Statistics classifies households with incomes in 1974 of less than K£420 per annum as 'lower income'; those with incomes between K£420 and K£1500 as 'middle income'; and households with over K£1500 as 'upper income'. Using the same classifications for the country-wide distribution of income in 1976 shown in Table 6.2, the first eight deciles fall in the lower-income group and the ninth falls in the middle-income group, only the tenth decile belongs to the upper-income group.

Killick[20] has further disaggregated these data and shown that, while those in the tenth decile receive 56 per cent of the income, the top half of this group share 44 per cent. His estimate of the Gini coefficient for the distribution of per capita income is 0.6, which is at the upper end of the scale of values generally obtained for this coefficient from income distributions.

In 1974–5, 67 per cent of all smallholder rural households, who make up the great majority of all Kenyans, received incomes of less than K£200 per annum. By treating children under fifteen years as equivalent to one-half of an adult, it has been further estimated that almost 45 per cent of all adult equivalents receive an income of less than K£25 and only one out of four rural families have annual incomes exceeding K£50 per adult equivalent. It is this 45 per cent of the population who have been called the 'rural poor', while about the same proportion of the population, with incomes between K£25 and K£75 per adult equivalent, are identified as 'the moderately endowed class'.[21] More recently, the current Development Plan identifies the most serious development problem as that of the working poor which comprises 49 per cent of the nation's total population. The groups include nearly all of the rural landless and squatters and pastoralists and nearly 40 per cent of small farm families.

It is no wonder that the Integrated Rural Survey revealed such a low accumulation of furniture items compared with the average number of household members. As household incomes rise, the number of basic items of furniture increases in the homestead. This does not necessarily mean, however, that furniture items per capita increase because of the high correlation between household income and household size.

There are no comparable estimates for the urban sector although the author's Nairobi informal-sector survey collected information about household incomes and a profile of household membership. It is estimated that 94 per cent of proprietors in Nairobi's informal sector are males with an average age of thirty-five years, suggesting that the great majority of respondents are household heads. Estimates suggest that 41 per cent of these households receive a total income of less than K₤300 per annum, which is perhaps equivalent to the K₤200 per annum in the 1974 rural survey. Furthermore, 22 per cent of all adult equivalents receive an income of K₤32 per annum, which may be compared with the K₤25 per adult equivalent in the rural survey.[22]

It appears that a large proportion of the urban population have low incomes, comparable with those of the rural poor. It is not unreasonable to assume that their holdings of furniture items are as sparse as those of the rural population. Therefore, we should expect that a policy of income redistribution that raised the incomes of these poverty groups would lead to a major expansion in the demand for items of basic furniture in both rural and urban areas.

We now turn to an examination of the pattern of furniture purchases by urban households in various income classes based on a special survey undertaken for this project. As part of its National Integrated Sample Survey Programme, Kenya's Central Bureau of Statistics has been conducting research with a random sample of households, addressing questions on the various characteristics of the labour force. In Nairobi the population has been stratified by areas of the city to reflect income classes, and a total of 1224 respondent households were visited once a month over a twelve-month period. Estimates of total household incomes were made each month, including wages and salaries, housing allowances and net incomes for those in self-employment. Average household incomes for the four monthly visits that were already computerised and analysed by November 1978 were supplied to the author for use in this project. The four observations on income were spread over the period November 1977 to July 1978 and are, therefore,

TABLE 6.2   Estimated income distribution in Kenya (1976)

Decile (1)	% share of total income (GDP) (2)	Per capita income K£ (3)	Household income K£ (4)
1st	1.8	16.5	79
2nd	2.0	18.3	88
3rd	2.6	23.9	114
4th	3.6	33.1	158
5th	4.0	36.7	175
6th	4.5	41.3	197
7th	5.2	47.8	228
8th	8.3	76.2	364
9th	11.7	107.4	512
10th	56.3	517.0	2 466

Source. Col. (2) from T. Killick, *Strengthening Kenya's Development Strategy: Opportunities and Constraints*, working paper no. 239 (Nairobi: Institute of Development Studies, 1976) p. 11. Cols (3) and (4) were taken from Kenya Development Plan 1979–83 estimates of 1976 total population and implicit assumption of 4.77 persons per household (Kenya, *Development Plan 1979–1983* (Nairobi: Government Printer, 1979)) p. 35. While household size rises with incomes, especially for the lower-income groups in rural areas, demographic trends suggest family size falls as incomes rise, particularly in urban centres. With no detailed information on household size by population decile the average of 4.77 persons was used to derive household incomes. Shares of income to each decile are assumed not unreasonably, to have remained unchanged between 1969 and 1976.

reasonably representative of average monthly income during the year.

Meanwhile, during the July–August 1978 round of the Nairobi survey a special module questionnaire addressed to furniture expenditures was administered by the enumerators. An attempt was made to gauge total household expenditure on various items of furniture during the previous one year. In addition, respondents were asked to identify the sources of their purchases, divided between the formal and informal sectors. The expenditures were then matched against the household income data that were independently supplied by the Bureau.

Items of furniture purchased from one of four possible sources were classified as being manufactured by the informal sector. The sources were: workshop in a temporary structure or in the open air; self-construction by a household member; a street hawker or informal-sector retail outlet; and a Nairobi City Council market stall. Formal-sector purchases were identified as those made from a small workshop in a permanent building, which would incorporate formal-sector enterprises in size-classes 1–19; and purchases directly from a factory or from a retail shop housed in a permanent building.[23]

A total of 321 households claimed to have made purchases of at least one item of furniture during the previous year and these formed the basis of our subsequent analysis. Households in the survey with monthly incomes of less than K£40 constituted 46 per cent of the sub-sample of purchasing respondents and are classified as 'low'; those with incomes between K£40 and K£75 made up 34 per cent of our sub-sample and are called 'medium'; those households with incomes exceeding K£75 per month are called 'high' incomes and they made up 20 per cent of our sub-sample.[24]

Table 6.3 presents the average prices of the various furniture items, which are indicators of cost of production and quality differences.

While for some items the number of observations is too small to draw general conclusions, for the most important basic items, such as tables, chairs and beds, average formal-sector prices are significantly higher than the corresponding informal-sector prices.

One of the major hypotheses of the project is tested in Table 6.4. Do low-income households show a greater tendency to buy from the labour-intensive informal sector, and is it generally the case that higher-income households buy superior-quality furniture from the less labour-intensive formal sector?

It is evident from Table 6.4 that the low-income households have a greater propensity to buy from informal-sector producers. For example, of the fifty-nine low-income households who had purchased a table, 81 per cent had bought an informal-sector product while only 19 per cent

TABLE 6.3   Average prices of items of furniture by sector of manufacture

Item	Average price per unit (Kenyan shillings)	
	Informal sector	Formal sector
Tables	105 (95)	150 (53)
Chairs	34 (126)	70 (38)
Beds	169 (45)	458 (35)
Cupboards	216 (29)	367 (18)
Wooden suitcases (boxes)	73 (28)	96 (12)
Shelves	86 (10)	180 (2)
Sofa sets	808 (6)	1 495 (47)

Note: Figures in parentheses are the number of observations for each item.
Source. Project Sample Survey.

TABLE 6.4 Average prices and number of buyers of informal- and formal-sector furniture, by income group

Item	Income class	Average price (Kenyan shillings)		Number of buyers	
		Informal	Formal	Informal	Formal
Tables	Low (L)	88	140	48	11
	Medium (M)	135	155	28	22
	High (H)	114	214	11	17
Chairs	L	30	45	70	13
	M	39	51	32	10
	H	33	113	12	13
Beds	L	167	229	23	10
	M	181	237	15	12
	H	180	874	5	11
Cupboards	L	204	200	10	1
	M	217	740	9	9
	H	175	434	3	6
Boxes	L	74	105	17	4
	M	81	69	7	4
	H	60	116	1	4
Shelves	L	56	–	7	0
	M	–	180	0	2
	H	155	–	3	0
Sofa sets	L	962	1 180	4	4
	M	700	1 463	1	22
	H	300	1 483	1	19

Source. Project Sample Survey.

had bought a formal-sector table. This pattern is established for almost all the items in the table.

Where low-income households buy formal-sector items, the average prices are invariably higher than the prices of the corresponding informal-sector product. Judging from these prices and comparing them with observations from visits to formal-sector establishments, these items have been purchased from the smaller-scale workshops. And, in the case of beds, the average formal-sector prices paid by the low- and middle-income groups closely correspond with the prices quoted for metal beds by the largest firm in the industry.

The propensity of the middle-income group to buy from the formal sector is higher than that of the lower-income households. Given their greater incomes, they purchase a higher-quality product than the less well-off families, judging from the higher prices they pay.

For the high-income families the pattern is also clear. In general they prefer to buy formal-sector furniture and in so doing they pay much higher prices than either those in the same income group buying from the informal sector or those in the low-income class buying formal-sector furniture.

Given the highly skewed distribution of income in Kenya, particularly in Nairobi, it is unfortunate that we failed in the survey to find many of the very high-income households who had bought furniture.[25] On the other hand, many of the executive and professional classes, both expatriate and Kenyan, do not buy furniture themselves but are provided with furnished accommodation by their employers. We observed earlier that firms in the 50–99 size-class who manufacture expensively upholstered and hand-carved furniture rely on these kinds of buyers. Those individuals buying such furniture, judging by the prices, also come from the class with very high incomes.[26]

A general pattern emerges from Table 6.4. The poor households in Nairobi, and by inference in other urban areas and in rural Kenya also, buy low-quality, relatively inexpensive furniture from the very labour-intensive informal sector. As incomes rise, preferences move in favour of more expensive and middle-quality furniture, produced in the small-scale workshops and large-size factories. Both kinds of producers are less labour-intensive than informal-sector producers but they are still very labour-intensive judged against other branches of industry. And both use a very small amount of imported raw materials.

At the top of the income scale, among the prosperous business and professional elite, are found the buyers of the luxury items of furniture. These are the very expensive, highly skill- and import-intensive products which have been identified with the firms in the 50–99 size-class, which exhibit well above average levels of labour productivity.

6.4 The Employment Effects of an Income Redistribution

Our profile of income distribution in Kenya has shown that 80 per cent of the population fall in the 'lower-income' group, as classified by the Central Bureau of Statistics. The government estimates that 49 per cent of the population is made up of the 'working poor', consisting of households with an annual income of less than K£150 who are unlikely to satisfy some of their basic needs. An earlier discussion drew attention to the plight of the two-thirds of rural smallholders with incomes of less than K£200, whose stocks of household furniture appear inadequate.

While poverty alleviation is the expressed goal of the government, no major static redistribution of income is envisaged since the present scope for increasing tax rates is somewhat limited. The rates of direct taxation of income now in force are progressive and for persons in the upper-income brackets they are relatively high.[27]

What the Plan envisages is a redistribution of the growth in national income through an improvement in agricultural productivity and market efficiency. The plan period is thus likely to bring about a redistribution of income within agriculture to the benefit of smallholders and pastoralists in low-income areas. For the working poor in the urban centres, employment opportunities are to be expanded and, through the use of various policy instruments, the productivity of their efforts, and in turn their incomes, will rise.

In the light of these statements, it would be unrealistic for this study to make any assumptions about major redistributions of income. In the following simulations it is assumed that future increases in income in the economy are redistributed in various ways to the bottom eight or nine deciles in the population. The likely impact on expenditure on furniture and changes in the industry's level of employment are then examined.

In the following simulations only the growth of monetary income is considered as amenable to redistribution by the government. In addition, self-produced furniture consumption will be very small and it is from monetary income that the vast majority of furniture is bought. Non-monetary or semi-monetary income in Kenya is essentially subsistence or own-produced goods. A national survey in 1963 generated data on the volume of production, marketed output and own consumption. This survey formed the base for subsequent estimates of non-monetary income which are made each year from changing population estimates and weather conditions. Home-produced consumption is valued at local market prices each year. Therefore, monetary income is monetised or marketed output and made up 81.4 per cent of total GDP in 1977.

A 7.1 per cent annual growth in monetary income is predicted by the Development Plan for the 1979–83 period. Monetary GDP at factor cost will have risen from K£1003 million in 1976 by K£623 million in constant prices in 1983. This represents an over-all increase in real income of 61.8 per cent.

Our first simulation assumes no changes in the shares of total income received by the three income groups. The bottom 80 per cent of the population claim 32 per cent of the total change in monetary income, the middle-income population of 10 per cent share 11.7 per cent of the gain, while the top 10 per cent retain 56.3 per cent of the increase.

From a 1974 survey, the Central Bureau of Statistics estimated that the following proportions of annual income are spent on furniture by the different income groups:

Lower income (<K₤420):	4.6 %
Middle income (K₤420–₤1500):	5.6 %
Upper income (K₤1500 +):	3.7 %

From insights gained from our expenditure survey, it is inferred that the lowest-income group buy the bulk of their furniture from the rural and urban informal sectors and firms in the 1–4 size-class; the middle-income group purchase from the informal sector and from workshops in the < 50 size-class of firms; and the richest group buy the highest-quality furniture from the 50–99 size-class of firms and medium-quality items from firms in the 100 + size-class.

While some purchases may be made by the poorest households from the largest-sized firms, for example, some metal beds, the great majority of Kenya's lowest-income group reside in the rural areas and conform to these purchasing patterns. Equally, the richest group of consumers buy odd items from the informal sector but these form a negligible part of their total purchases.

Without being able to determine exactly the relative proportions of expenditures among the various size-classes of firms, it is assumed that an income group spreads its purchases equally over the kinds of firms from which it buys. For example, the lowest-income group buys half of its furniture from the informal sector and half from the 1–4 size-class, and so on. And from the informal-sector survey and the 1972 Census of Industrial Production it is possible to estimate the average ratio of value added to value of sales for the different size-class of firms.[28]

While simulation 1 assumes no redistribution of income shares, simulation 2 assumes that the richest 10 per cent of the population only receive enough of the growth in income to maintain their 1976 level of per capita income. Of the additional income available for distribution in 1983, simulation 2 assumes that one-third goes to those in the ninth decile and two-thirds goes to the poorest 80 per cent. Simulation 3 assumes the rich and middle-income groups maintain their 1976 levels of per capita income and all of the remainder of the growth in income is distributed to the poorest 80 per cent.

Table 6.5 illustrates that the relatively 'mild' income redistribution of simulation 2 stimulates an additional 5471 jobs over the increase in employment that would have occurred with no change in income shares as in simulation 1. Meanwhile the 'radical' redistribution of income of

TABLE 6.5 Estimated change in furniture industry's employment by 1983 over 1976 level

Income group	% share of increase in income in 1983	Additional expenditures (K£m)	Additional value added (K£m)	Additional employment
Simulation 1				
Low	32.0	9.126	3.65	9 456
Middle	11.7	4.062	1.52	2 468
Upper	56.3	12.914	4.78	4 906
Total	100.0	26.102	9.95	16 830
Simulation 2				
Low	39.0	14.333	5.73	14 853
Middle	16.8	8.725	3.27	5 303
Upper	44.2	5.647	2.09	2 145
Total	100.0	28.705	11.09	22 301
Simulation 3				
Low	46.6	20.039	8.02	20 765
Middle	9.2	1.777	0.67	1 079
Upper	44.2	5.647	2.09	2 145
Total	100.0	27.463	10.78	23 989

simulation 3 would add an additional 7159 jobs to the increase without income redistribution.

We estimated total employment in the furniture industry at roughly 30 000 in the 1972–4 period. Accordingly, our simulations represent annual rates of increase over the ten-year period of 4.6 per cent, 5.7 per cent and 6.1 per cent respectively (Table 6.6).

TABLE 6.6 Projected employment growth in the furniture industry under alternative simulations

	Furniture expenditure patterns			
	as per CBS		with 7% assumption for higher group	
	Additional employment	Annual rate of growth %	Additional employment	Annual rate of growth %
Simulation 1	16 830	4.6	21 205	5.4
Simulation 2	22 301	5.7	24 214	6.1
Simulation 3	23 989	6.1	25 902	6.4

The Central Bureau of Statistics (CBS) estimates that the proportion of income spent on furniture rises between the 'lower-' and 'middle-' income groups but falls back for the 'higher-' income group. This implies that as a household rises from the bottom income group to the highest group its income elasticity of demand for furniture would initially be in excess of one, but would ultimately become inelastic. While it is notoriously difficult to estimate such elasticities for durable goods, one study of Nigeria[29] suggests an expenditure elasticity of 4.95, where durables include furniture, household equipment and passenger cars.

The estimated 1976 levels of per capita income by group are given in Table 6.8 as K₤37, K₤107, and K₤517, respectively. The CBS furniture expenditure estimates of 4.6 per cent, 5.6 per cent and 3.7 per cent of income of the three groups implies an expenditure elasticity of 1.34 between the lower- and middle-income groups, and 0.57 between the middle- and upper-income groups. An additional simulation was carried out on the assumption that the upper-income group spend 7 per cent of their income on furniture. That is, at existing income levels, the expenditure elasticity between the middle- and upper-income groups becomes 1.32.

As a result the increase in employment in simulation 1 became 21 205, with no redistribution in the shares of the predicted growth in the economy between 1976 and 1983. This represents an annual growth in employment of 5.4 per cent, while simulation 2 represents a 6.1 per cent growth rate and simulation 3 implies an increase of 6.4 per cent per annum. This amendment in no way alters the conclusions that follow from the initial simulations. A policy of income redistribution through growth will accelerate the rate of increase in employment in Kenya's furniture industry.

Using the CBS estimated expenditure patterns by income class, Table 6.7 illustrates the number of new job opportunities and the average incomes generated for their holders.[30]

Table 6.7 shows that most of these additional jobs would be created in the rural and urban informal sectors, where the average earnings are lowest. The more severe the income redistribution taking place the greater would be the growth of lower-income employment.

The urban informal-sector survey would lead us to expect the increased incomes generated to exceed the average formal-sector wage in the case of the owners of such businesses and to lie somewhere between the rural and urban legal minimum wage levels in the case of employees.

TABLE 6.7   Projected employment created, by average incomes per job, under
alternative simulations

| Simulation | New jobs by average remuneration | | |
	K£437	*K£530*	*K£641*
1	9 546	2 468	4 906
2	14 853	5 303	2 145
3	20 765	1 079	2 145

Note: These are short-term projections which assume minimal changes in technology.

As a result of these increases in incomes, employment and output of low-quality furniture, the goal of satisfying the basic needs of the poorest sections of Kenya's population would be clearly enhanced. From the urban informal-sector survey it was established that the vast majority of business operators are themselves the owners of the enterprises, and there is no reason to believe the situation in the rural sector is any different. Therefore, any additional incomes arising in the sector from the assumed income redistributions will accrue to the owners and employees in the sector and will not be appropriated by absentee owners in other parts of the economy.

The effect of the various income redistributions with respect to the goal of alleviating poverty is illustrated in Table 6.8.

With no income redistribution assumed, in simulation 1, income inequalities widen and any improvement in the plight of the poorest 80

TABLE 6.8   Per capita and household incomes under the various
assumed schemes for income redistribution in 1983, in 1976 prices

| Per capita income (*K£*) by group | Estimated 1976 level | Simulations | | |
		1	*2*	*3*
Lower	37	44	52	60
Middle	107	129	176	107
Upper	517	623	517	517
Household income (K£) by group				
Lower	175	211	248	286
Middle	512	618	840	512
Upper	2 466	2 973	2 466	2 466

Note: Income is measured here as total GDP at factor cost, including the income accruing in the semi-monetary economy. Average household size is assumed to be 4.77. Total GDP is included here to give the reader a full picture of changes in total real incomes of the three groups under various simulations.

per cent is negligible. On the other hand, under simulations 2 and 3 the extent of income inequality is sharply lessened. In simulation 3, while the richest 20 per cent of Kenyans maintain their real incomes at 1976 levels, the poorest 80 per cent of the population realise increases of over 60 per cent in real per capita and household incomes.

Meanwhile, the manner in which this income is shared among the bottom 80 per cent will, of course, determine the numbers still falling into the category of the working poor, with household incomes of less than K£150 per annum.

6.5 Conclusions

Earlier macro-level studies of the secondary employment and income effects of an income redistribution ignored a key element in the income distribution–technology–employment hypothesis. No account was taken of the diversity of factor proportions within each industry, which is reflected in the size distribution of firms. The recent study by Thirsk[31] allowed the intra-industry composition of firm size to vary in response to a simulated income redistribution. By assuming that the poor spend their increased incomes on the products of the relatively labour-intensive small firms, his findings reverse the earlier pessimistic conclusions and show that for every dollar transferred the incomes of the poor ultimately rise by as much as two or three dollars.

This case-study has been addressed at the level of a particular product group. It has explored both the kinds of technologies employed by firms of different sizes in the furniture industry in Kenya and the expenditure patterns of consumers across income groups. An expenditure survey has been used to reveal that the poor buy furniture predominantly from the smaller, more labour-intensive firms in the industry.

The study examined in this chapter has shown that a major increase in employment in the furniture industry would be a consequence of this policy of income redistribution through growth. Such an expansion would create quite respectable income levels for the newly employed and go a long way to alleviate existing deficiencies in stocks of furniture among the poorest Kenyan households.

Notes

1. ILO Employment Planning Expert, Department of Statistics and Research, Ministry of Finance, Cyprus. The author wants to thank Tony Killick, A.S.

Bhalla, Wouter van Ginneken, Les Morrison and Maurice Thorn for their valuable comments on earlier drafts. This chapter is a short version of an article that appeared in the *Journal of Development Studies* (July 1981).

2. The government's goals for the 1979–83 Development Plan are expressed in the statement that 'the provision of basic needs will play the key role in the over-all development strategy of the Development Plan whose principal theme is the alleviation of poverty'. See *Development Plan 1979–1983* (Nairobi: Government Printer, 1979).

3. M. Sheridan (ed.), *The Furnisher's Encyclopaedia* (London: National Trade Press, 1955).

4. Central Bureau of Statistics, 'Non-farm Activities in Rural Kenyan Households', *Social Perspectives* (Nairobi), no. 2 (1977).

5. In addition, 30 per cent of the businesses operating in the four major government-sponsored Rural Industrial Development Centres are engaged in furniture. See I. Livingstone, *Kenya's Rural Industrial Development Programme: An Evaluation of Experience and Proposals for Action*, working paper no. 210 (Nairobi: Institute of Development Studies, University of Nairobi, 1975).

6. The 'modern' sector is defined as 'the entire urban sector, the entire public sector, large-scale farms and other large-scale enterprises such as sawmills and mines, located outside towns'. However, the urban informal sector is excluded from this definition and remains largely unenumerated, as is true of small-scale activities in rural areas. See Central Bureau of Statistics, *Statistical Abstract 1978*; and O. Agunda, *Data Collection in the Urban Informal Sector in Kenya*, occasional paper no. 25 (Nairobi: Institute of Development Studies, University of Nairobi, 1978).

7. W. J. House, 'Market Structure and Industry Performance: The Case of Kenya', *Oxford Economic Papers*, vol. 25, no. 3 (1973).

8. The Nairobi informal-sector survey was carried out by the author in mid-1977. Enterprises were classified as belonging to the sector if they operated out of a temporary structure or from no structure at all. Activities housed in a concrete, brick or cement block structure were excluded from the sampling frame. See W. J. House, *Nairobi's Informal Sector: A Reservoir of Dynamic Entrepreneurs or a Residual Pool of Surplus Labour*, working paper no. 347 (Nairobi: Institute of Development Studies, University of Nairobi, 1978). Of the 578 respondents, 93 were randomly selected manufacturers of wooden furniture and, from Inukai's description of rural artisans, they are representative of the small-scale rural carpenter as well. See I. Inukai, *Rural Industrialisation: A Country Study, Kenya* (Nairobi: Department of Economics, 1972; mimeographed).

9. For a more complete description see W. J. House, *Technological Choice, Employment Generation, Income Distribution and Consumer Demand: The Case of Furniture Making in Kenya* (Geneva: ILO, 1980; mimeographed World Employment Programme research working paper; restricted).

10. Livingstone, *Kenya's Rural Industrial Development Programme*.

11. F. Stewart, *Technology and Underdevelopment* (London: Macmillan, 1977).

12. Average value added per worker in formal-sector furniture in 1972 was K£1008 and for all manufacturing was K£1548, both quoted in 1977 prices. The nature of the labour intensity of the industry is also apparent from

comparisons with the UK industry. For the group of industries, glass, cement, paper and printing and furniture, output per person in 1976 was K£4775, which is nearly five times greater than that for furniture in Kenya. See Central Statistical Office, *Annual Abstract of Statistics, 1979.*

13. An executive of the Marketing Society of Kenya assured the author that the total amount of advertising expenditure by the industry did not exceed K£10 000 in 1978. By comparison the Coca Cola company spent almost K£100 000 during the year on advertising its product alone.

14. R. Nelson, 'A Diffusion Model of International Productivity Differences in Manufacturing Industry', *American Economic Review* (Dec 1968).

15. P. Henning and W. J. House, 'The Problem of Slow Employment Growth in the Manufacturing Sector of Less Developed Countries', *Journal of Eastern African Research and Development*, vol. 3, no. 2 (1976).

16. The net income accruing to informal-sector furniture manufacturers was estimated indirectly, from answers to questions about value of sales, wages, raw materials, rent and depreciation for the reference week preceding the interview. These data were inflated by a factor of fifty-two to derive annual net income of the business.

17. See House, *Nairobi's Informal Sector*, pp. 14–15.

18. The capital includes the estimated replacement value of tools and equipment, stocks of raw materials and finished goods and the monthly building rent, capitalised over twenty-five years. Output is the estimated annual value added of the business.

19. H. Chenery, M. Ahluwalia, C. Bell, J. Duloy, and R. Jolly, *Redistribution with Growth* (London: Oxford University Press, 1974).

20. T. Killick, *Strengthening Kenya's Development Strategy: Opportunities and Constraints*, discussion paper no. 239 (University of Nairobi, 1976).

21. L. Smith, *Low-income Smallholding Marketing and Consumer Patterns*, FAO Marketing Development Project (Nairobi, 1978: mimeographed).

22. While these comparisons allow for price inflation between 1974 and 1977, no allowance is made for cost-of-living differences between the rural and urban areas, for which we have no information.

23. While it could be that formal-sector retailers sell items made in the informal sector, the survey results indicate that such marketing channels are of limited significance for informal producers.

24. The Central Bureau of Statistics currently calls 'lower-income' households those below K£35 per month, 'middle-income' between K£35 and K£125 and 'upper-income' over K£125. They suggest 25 per cent of households in Nairobi fall in the 'lower' group, 50 per cent in the 'middle' group and 25 per cent in the 'upper' class. In our survey only 10 per cent of households had incomes in excess of K£125 which, given the small number in the sample, would have been too few to analyse. It would appear that, if the Bureau's estimates are correct, our survey overrepresents the low-income group, or respondents have understated their incomes. These anomalies do not present any major problems since the main task is to gain a general impression of expenditure patterns for the different income groups.

25. Only 3 per cent of the respondents had household incomes in excess of K£250 per month.

26. During personal interviews conducted by the author it was discovered that

both the University of Nairobi and the major government departments buy furniture for their professional staff from the largest producer in the industry.

27. *Development Plan 1979–1983*, p. 110. The highest rate of tax is 65 per cent, levied on incomes exceeding K±9600.

28. These ratios were: informal sector, 0.528; size 1–4, 0.263; 5–19, 0.359; 20–49, 0.350; 50–99, 0.350; 100 +, 0.393.

29. R. Weiskoff, 'Demand Elasticities for a Developing Economy: An International Comparison of Consumption Patterns', in H. Chenery (ed.), *Studies in Development Planning* (Cambridge, Mass.: Harvard University Press, 1971).

30. Average income per job is obtained from the assumed expenditure patterns of the income recipients and the average earnings of workers by size-class of firms, in Table 6.1. For example, the average income per new job of K±437 is obtained by assuming one-half of new employees receive the average wage of K±528 of the 1–4 size-class of firms; and of the remaining 50 per cent of new workers one-third are entrepreneurs in the informal sector receiving K±707 and two-thirds are their employees receiving K±166 per annum. The other average income levels were generated in a similar manner. It is impossible to distinguish between rural and urban informal employment generation attributable to any of the income simulations.

31. W. R. Thirsk, 'Aggregation Bias and the Sensitivity of Income Distribution to Changes in the Composition of Demand: The Case of Colombia', *Journal of Development Studies* (Oct 1979).

7 Technology, Products and Income Distribution: The Soap Market in Barbados

Jeffrey James[1]

Most economists 'accept unquestioningly the consumer's judgement of what is best for him, his tastes as the outcome of that judgement, and his market behaviour as the reflection of his tastes'.[2] The basis of this view is the traditional theory of demand, particularly its revealed preference variant, which *assumes* that the consumer is always in equilibrium, maximising satisfaction from the goods that he purchases. Not surprisingly, therefore, economists have paid little or no attention to the 'quality' of decision-making by consumers. There is a vast literature on production efficiency; almost none on consumption efficiency. Recently, however, demand theory has produced a new approach which does permit analysis of this question.

Associated mainly with Lancaster, the new theory views the demand for goods as a derived demand for the characteristics embodied in them.[3] In this chapter it is shown that the new theory can be used to throw considerable light on the 'efficiency' of consumption processes in developing countries. Moreover, because products change rapidly over time, altering the choice available to consumers, we argue that the static analysis of choice from existing products should be complemented by a more dynamic approach which considers the effects of product changes on the different groups in these societies.

The chapter is organised as follows. The first part sets out the theoretical basis of the study in both its static and dynamic dimensions. This is followed by an application of the theoretical framework to data collected during fieldwork in Barbados. The final part of the chapter deals with policy implications of the findings.

7.1 Theoretical Framework

Static analysis: the choice from existing products

The numerous requirements for effective decision-making by the consumer in the context of the product characteristics approach can usefully be approached in terms of a simple figure (Figure 7.1).

Figure 7.1 shows the simplest case of two goods *A* and *B* embodying characteristics 1 and 2 in differing proportions. *II* is the indifference map for a single consumer. It is assumed that characteristics 1 and 2 are objective, that is, they are in principle capable of being measured. For example, the characteristics 1 and 2 may represent certain nutrients of foods, weights or lengths of goods, etc. Goods *A* and *B* can then be objectively located in the two-dimensional space representing these characteristics.

If the consumer is to reach *R*, his point of maximum satisfaction, a number of requirements need to be satisfied.

Accurate perception of product characteristics. Perhaps the most obvious requirement is that the consumer be able to locate the vectors *A* and *B* in the characteristics space. That is, the individual needs to know that the goods exist as well as how they map into characteristics (in

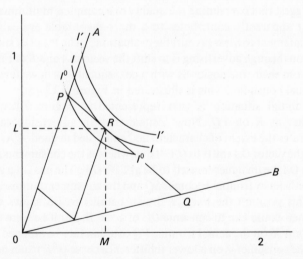

FIGURE 7.1   The requirements for effective decision-making by the consumer

addition, of course, his preferences for the set of characteristics have to be known).

The sources of information regarding the objective characteristics of commodities are many and varied. In some cases consumers can themselves discern the required information by using sensory perceptors (such as touch, smell, visual images, etc.) but in many others they have to rely on alternative sources. Clearly, the scope for a divergence between the actual characteristics of goods and that perceived by individuals is far greater where complex technical information is involved but even the visual inspection of simple objective characteristics can sometimes be a complex process. For example, a number of studies have shown that characteristics of objects, such as size, which are independent of their value can nonetheless be somehow distorted as a function of that value.[4] Other studies have shown that 'individuals modify their judgements of perceived objects, such as lengths of lines, to conform to the judgement of a social group, or the opinions expressed by other members of the group'.[5] It has been demonstrated furthermore that need and perception can interact, as when the brightness of pictures of food varies with the extent of the hunger of the observers.[6]

It is often unfortunately the case that in the very situations in which the need for reliable information is greatest (namely, with goods whose characteristics are very difficult to measure), it is also least likely to be forthcoming. Thus, 'while business ethics and consumerism may clearly direct the promoter to make realistic claims for his product, the results here suggest that overstating the quality of a complex, multidimensional product apparently contributes to a more favourable evaluation and understatement to a less favourable evaluation'.[7] The result of such over-valuation through advertising is to shift the vectors A and B of Figure 7.1 away from their true positions with a consequent loss in welfare for the individual consumer. This is illustrated in Figure 7.2.[8]

The initial situation is that represented in Figure 7.1 with the consumer at R on PQ. Now assume that an advertising campaign exaggerates the extent of characteristic 2 embodied in good A. As a result of this the vector OA shifts to OA'. Thus, whereas the consumer is actually at P on OA, he imagines himself to be at S on OA'. The imaginary budget line or efficiency frontier becomes SQ and the consumer optimises at T on it. In this position the line CT drawn parallel to OS shows that the consumer would like to consume OC of good B. But if he were to do so only point X on the actual frontier PQ could in fact be attained. At this point the consumer is on a lower indifference curve (I^0I^0) than he would have been (II) given accurate information regarding characteristic 2. One

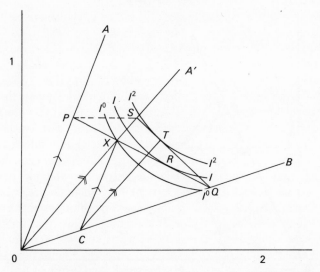

FIGURE 7.2   The effect of over-valuation through advertising

corollary of this analysis is that misinformed consumers, when presented with full information, are likely to alter their purchasing patterns so as to reach a higher level of welfare. A recent example of this is the shift away from phosphate-intensive detergents following the revelation of their ecologically harmful effects.

Divisibility.   If the consumer in Figure 7.1 is to reach his most preferred point R, it is plain that a condition of divisibility is required.[9] If, for example, goods A and B are perfectly indivisible, the best that the consumer can do is to move to P on the lower indifference curve I^0I^0. Partial indivisibility will enable the consumer to move closer to R and perhaps even to reach it.

Accounting skills.   Even with perfect information and divisibility it is still not necessarily the case that an individual is consuming efficiently. One reason is that managerial or accounting skills are needed to ensure that goods are combined in the desired proportions.

Figure 7.3 shows that if the consumer is to reach R, his budget has to be allocated to the two goods in precisely the right proportions, namely, PR/PQ to B and RQ/PQ to A. Should even the slightest computational error arise the consumer will be forced on to a lower indifference curve.

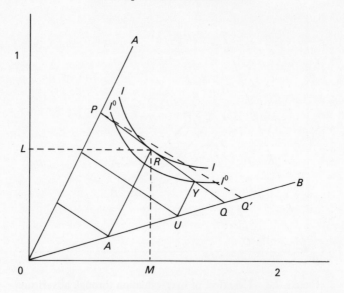

FIGURE 7.3   The need for accounting skills

If, for example, slightly too much is spent on good *B* (so that the proportion spent on this good rises to *PY/PQ*) the consumer will end up at *Y* on the lower indifference curve I^0I^0.

Not only may consumers commit computational errors, they may also fail to make any computations at all. Most commodities can be purchased in different sizes with varying unit prices. If these are not calculated by consumers or provided by the manufacturers, inefficiency is bound to arise. If in Figure 7.3, for example, the unit price rises with larger-sized versions of good *B*, those buying the good in small units in effect shift the efficiency frontier outwards (to say *PQ'*).

Product use skills. Thus far it has implicitly been assumed that if a commodity embodies certain characteristics then its mere purchase is sufficient to ensure that the embodied quantities are actually *attained*, that is, that at point *R* in Figure 7.3, *OL* of characteristic 1 and *OM* of characteristic 2 are actually received. But there is no reason why this need always or even generally be true.

In the modern world many goods, if they are to be used properly, require a host of consumption skills. In many cases instructions (sometimes complex) have to be followed for efficient product use. In addition, if the maximum amounts of characteristics embodied in products are to be realised, the latter require to be used in a context where necessary complementary resources exist (e.g. hot water for many soap powders, freezers for frozen foods). Product use in the absence of complementary factors will lead to a reduction in the characteristics obtained from the total potentially available.

Problems of product use are likely to be greatest in low-income countries particularly in relation to new products.[10]

The limitations of static analysis

The entire analysis so far is static – it takes as given the types of products available, their prices, sizes, etc. But the types of goods available today (in terms of the characteristics embodied in them) are the outcome of yesterday's configuration of circumstances and what is available today has an impact on the distribution of income.

In the Lancaster approach technical change may be conceptualised as follows.[11] Assume a world of (i) commodities, and (ii) characteristics. Let x_{ij} be the amount of the ith characteristic obtained from the jth good (where $i = 1, \ldots, b$ and $j = 1, \ldots, a$). Let X_j be the amount of the jth consumers' good. It is then possible to define an input–output coefficient y_{ij} as the amount of the ith characteristic derived from a unit of the jth good. That is, $x_{ij} = y_{ij} X_j$.

Since there are ab input-coefficients, the consumption technology can be represented in usual matrix form. Technical change comprises alterations in one or more cells of the matrix or additions to it. The distributional impact of technical change can be illustrated as in Figure 7.4.

This figure portrays a situation of two consumers with identical incomes and differing tastes represented by the indifference curves II, and $I'I'$ for individuals 1 and 2 respectively. The initial position in both parts of Figure 7.4 has two goods x and y with the individuals 1 and 2 consuming at A and B respectively. In the first case (Figure 7.4(i)) the new good Z extends to the right of good Y and extends the efficiency frontier from AB to AD. In this event all the gains from the introduction of the new good accrue to individual 2. In the second case (Figure 7.4(ii)), by contrast the new good extends the efficiency frontier from AB to DB. Here all the gains accrue to individual 1 and 2 does not benefit at all. Examples of goods embodying characteristics in vastly different proportions and meeting the same broad need are not difficult to find. Radios for instance can be designed that cost hundreds of pounds or, if specially

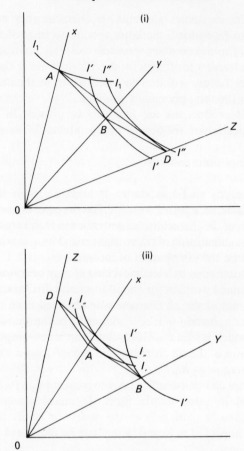

FIGURE 7.4 The distributional impact of technical change

designed for those with very low incomes, 'an entire unit can be made for just below 9c (US)'.[12]

The direction taken by the research effort in the design of products is determined, in the capitalist countries at least, by the money votes of consumers which thus tends systematically to bias the development of new goods in an inegalitarian direction. More particularly, products are designed for the majority in terms of money income but the minority in terms of the global distribution of individuals. This means that products tend to embody characteristics in proportions that accord with the preferences of those in rich rather than poor countries. It is this pattern that advocates of 'appropriate products' in developing countries seek to alter by the design of goods corresponding to the tastes of the majority of those living in the Third World.[13]

Today's goods may, in addition to their effect on *relative* gains and losses described above, also result in *absolute* losses for certain groups in the society.[14]

If the introduction of a new good merely extends the range of efficient choices available, then obviously no such absolute losses can occur. In many situations, however, the new good either totally displaces existing goods or leads to altered cost conditions for them. For example, the new and existing goods may compete in using the same scarce factor of production, and the production of the new good raises the cost of production of the old. Secondly, where there are economies of scale and diseconomies of small scale, so that if demand shifts away from existing to new products, costs of production of the former rise. Finally, where markets are dominated by monopolies or oligopolies, these firms may withdraw their old products to 'make room' for the new.

The way in which some consumers gain at the expense of others as a result of the introduction of a new good is demonstrated in Figure 7.5.

The initial efficiency frontier is given by *AB*. The introduction of the new good, *Z*, leads to a rise in the cost of production of *X* such that the new frontier is represented by *A'CB*. Whether or not consumers gain or

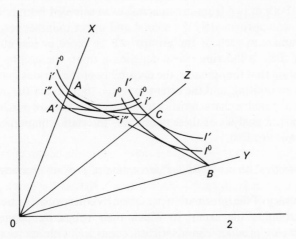

FIGURE 7.5 New products and absolute losses

lose from this change depends on their tastes. Those with indifference curves *II* will gain, while those with the *ii* map will lose absolutely.

7.2 A Case-study

This section describes the results of a case-study of the laundry soap and detergent market in Barbados. The study, which was begun in the middle of 1979, sought to determine, first, how well the various criteria for efficient consumption defined in the previous section were satisfied by different income groups and, secondly, how changes over time in the characteristics of laundry soap and detergents have affected the distribution of income.

Information on at least some of these questions had to be sought on the basis of actual interviews with consumers and could not be deduced merely from observation of market behaviour. For example, the observation that a particular consumer purchases a certain brand reveals nothing about his knowledge of its various characteristics, nor does it indicate whether he can purchase his desired brand in the preferred size (i.e. whether indivisibility is a problem). Interviews were conducted with approximately 330 households chosen on the basis of a quota sample with income as the major control.[15] Of the four more or less equal size sample groups, two were drawn from households falling into roughly the poorest quintile of the population (one group was drawn from rural areas and the other from urban locations). The 'medium' and 'rich' groups (both drawn from urban areas) were sampled from households falling into approximately the second and upper quintiles respectively.

Consumers in each of the groups will be more or less efficient in meeting their laundering needs depending upon the accuracy of the information that they possess, the divisibility of the various goods, their skills in accounting, and the extent to which they obtain the maximum amount of possible characteristics from the goods that are purchased. To the empirical analysis of these questions – the static component of our study – we turn first.

The 'subjective' versus 'objective' perception of product characteristics

The accuracy of the information possessed by consumers can be assessed by determining the extent to which their subjective perception of (measurable) product characteristics coincides with more objective measures.

The four products that were chosen for this assessment were all known to most of the respondents in Barbados since each has, at one time or another, enjoyed a substantial share of the laundry-cleaning market. Prior to the introduction of synthetic detergents (i.e. products containing surface-active agents other than soap) in Barbados in the 1960s, laundry soap in bar form was the dominant laundry cleaning product. Tide was the first synthetic detergent to capture a large part of the market from laundry soaps, only to be replaced in turn by Breeze and Drive, which currently account for some 60 and 30 per cent of the total market respectively. The prices of Breeze, Drive and Tide and the brand of laundry soap selected for the study are shown in Table 7.1.

Not surprisingly, in view of the fact that it is sold without any packaging, the laundry soap is by far the cheapest of the four brands. Tide is the most expensive, mainly because – unlike Breeze and Drive which are manufactured in Trinidad and are thus imported duty free into Barbados under Caribbean Common Market (CARICOM) arrangements – it is subject to a 30 per cent import duty.

As a first step in determining the subjective perception of these four products by the sample households, eleven characteristics of laundry-cleaning products were chosen. These (cheapness, whiteness of wash, popularity with people, attractiveness of scent, stain-removing ability, versatility in household use, effect on hands and fabrics, ease of lather, attractiveness of packaging and soil-removing ability) were the characteristics that the pilot survey had shown to be most important to consumers' choice. Each respondent was asked to rate the four products according to all eleven characteristics on a 1 to 7 semantic differential scale. With respect to soil-removing ability, for example, the scale takes the form shown in Table 7.2.

By averaging the ratings for the entire sample, the outcome is a 4×11 matrix of mean ratings shown in Table 7.3.

Unlike the simple figures with two products and two characteristics

TABLE 7.1   Unit prices – mid-1979

Brand	Price per ounce (Barbados cents)	Size
Blue Soap	5.0	8 oz
Breeze	12.5	16 oz
Drive	15.0	14 oz
Tide	18.7	20 oz

TABLE 7.2　The semantic differential scale

Soil-removing ability

1 -------	2 -------	3 ------	4 ------	5 ------	6 ---	7 --------
Extremely poor	Very poor	Slightly poor	Neither poor nor good	Slightly good	Very good	Extremely good

TABLE 7.3　Mean ratings for entire sample*

Product characteristics	Breeze	Blue Soap	Tide	Drive
Cheapness	4.7	3.6	5.3	5.4
Whitening ability	5.6	5.3	5.0	6.1
Popularity	6.2	3.9	3.6	5.7
Attractiveness of scent	5.8	3.0	5.3	5.7
Stain-removing ability	3.7	3.5	3.8	5.6
Household uses	5.7	3.4	5.2	5.6
Hand care	4.9	4.6	4.4	4.6
Ease of lather	6.3	3.5	5.3	5.9
Attractiveness of packaging	5.9	1.8	5.5	5.6
Soil-removing ability	5.8	4.9	5.2	6.3
Fabric care	5.5	5.2	5.1	5.0

* Where one or more ratings for a respondent were missing, his/her entire matrix of ratings was discarded. These were then ignored in the calculation of mean ratings for the entire sample and for the four income groups.

with which we were mainly concerned in the previous section, the data of Table 7.3 are multidimensional – there are four products and eleven characteristics. As a result, it is more difficult to discern the main perceptual differences between the four products. Recently, however, a new set of multivariate techniques have been developed (by mathematical psychologists) which enable these differences to be more clearly seen.

In common with techniques of factor analysis and principal components analysis, multidimensional scaling techniques facilitate interpretation of a set of complex data by reducing its dimensions. Obviously such simplification cannot be achieved without a cost; indeed, the greater the extent to which the dimensions are reduced, the greater will be the distortion of the original data. That is, the simplification is achieved at the expense of accuracy. Precisely where the balance between these two factors should be struck is a complex matter on which the interested reader can obtain some guidance from the technical literature.[16]

One multidimensional scaling model has achieved considerable popularity in marketing because it allows products and characteristics to be represented in the same (often two-dimensional) 'joint space'. Known as the point-vector model, it produces a configuration (in a number of dimensions chosen by the user) with points representing products (i.e. the columns of Table 7.3) and vectors representing characteristics (i.e. the rows of Table 7.3) such that projection of the points on the vectors provides the best linear approximation to the original data (Table 7.3). In other words it attempts to position the points (representing the columns of the matrix) and the vectors (representing the rows of the matrix) in such a way that the projection of the former on the latter provides the best (in a least-squares sense) approximation to the original data.

The operation of the model and its usefulness can be illustrated with an application to the data of Table 7.3 (for a full description of the analytics of the model and its applications the reader is again referred to the specialist literature).[17] Figure 7.6 shows, for example, a two-dimensional configuration with the four products, Breeze, Drive, Tide and Blue soap, as points and the eleven characteristics as vectors.

The configuration is interpreted by projection of the product points on the vectors (which provides the best linear approximation to Table 7.3 in this number of dimensions). For example, Drive is the most highly rated on stain removal, followed by Tide, Breeze and Blue Soap. In the case of hand care, by contrast, it is Breeze that is most highly rated, followed by Drive, Blue Soap and Tide. In general Figure 7.6 suggests that Drive is distinguished by its 'strength' (and negatively by its 'abrasiveness') while Breeze is thought to be mild/gentle in relation to the other brands. Tide has no particular image in the minds of consumers and Blue Soap receives the lowest ratings on most of the eleven characteristics.

In the same way configurations were also obtained for each income group separately. Though there were one or two differences between them,[18] they were basically very similar to Figure 7.6 and are therefore not reproduced.

We may now compare the perceptual summary contained in Figure 7.6 with a more objective evaluation of the characteristics where these are objectively measurable. Any disparity between the subjective and objective perceptions will then provide a measure of imperfect information (for the 'average consumer' of the sample).

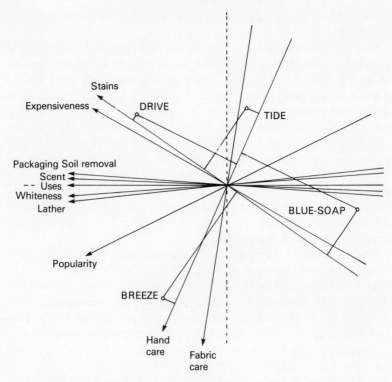

FIGURE 7.6　Joint-space configuration of four products and eleven
characteristics

Out of the eleven characteristics, only four – whiteness of wash, ease
of lather and soil- and stain-removing ability – were amenable to any
sort of objective test (fabric care over time can also be measured but
requires testing over a long period). On a washing test performed in a
laboratory the four products were rated on these four characteristics
with the results shown in Table 7.4.[19]

Because of the inherent difficulties of testing washing products these
results should be viewed with a great deal of caution. Apart from the low
ratings of Blue Soap, however, a fairly marked divergence between the
subjective and objective perceptions is indicated. In particular, the
image of the 'strength' of Drive and the 'mildness' of Breeze which was
revealed in Figure 7.6 is not (apart from whitening ability) confirmed by
the laboratory test. With respect to soil- and stain-removing ability, no

TABLE 7.4 Laboratory ratings*

	Breeze	Blue Soap	Tide	Drive
Ease of lather	7	2	4	7
Soil-removing ability	6	2	6	6
Stain-removing ability	6	2	6	6
Whiteness of wash	5	2	5	8

* The laboratory ranked the four products and we assigned scores of 8, 6, 4 and 2 for first, second, third and last rankings respectively.

difference at all between the two products (or between these products and Tide) was found to exist.

There are many factors (such as random errors in perception, difficulties in accurately comparing brands, etc.) that could account for this divergence of perceptions. But the most obvious source of the discrepant ratings is advertising that stresses the 'power' of Drive relative to other detergents and the fact that Tide is not advertised at all.[20]

Divisibility

Unlike the brands imported from the developed countries which are not normally packaged in sizes less than 16 ounces, those manufactured in the Caribbean, such as Breeze and Drive (which are produced by Lever Brothers in Trinidad), are packaged in small sizes of 7 or 8 ounces. None the less, as Table 7.5 shows, there was a big difference among the four income groups in their response to the question 'Do you sometimes wish to purchase in smaller quantity packages than are easily available?'

TABLE 7.5 Indivisibilities in consumption

	Urban poor	Rural poor	Urban medium	Urban rich
% unable to purchase in sufficiently small units*	52	64	40	3

* 3, 4 and 5 per cent of the urban medium, urban poor and rural poor groups were not asked this question since they were not using a detergent.

It would appear therefore that indivisibility is scarcely a problem for the rich group but affects the poor and middle-income consumers considerably. There are a number of possible reasons for this. First, the very poorest groups may wish to purchase in units even smaller than the small packages of Breeze and Drive, the brands used by the vast majority of all income groups. Secondly, the Lever Brothers plant in Trinidad had, during the year preceding the survey, suffered from industrial disputes with a consequent reduction in output. Since the most economical size package from their point of view is 16 ounces, they cut back production to a disproportionately great extent on the small sizes purchased mainly by the low-income groups. The final reason is that in periods of acute shortage, even the poor are forced to purchase brands manufactured outside the Caribbean, which, as noted above, are typically not packaged in sizes less than 16 ounces.

Accounting and product use skills

The survey revealed some interesting differences in attitudes to consumption, consumer skills and product usage among the four income groups. In particular, a contrast was found between the 'modern' practices and attitudes to consumption of the high-income households on the one hand, and the more 'traditional' orientation of the low-income groups on the other.

A good example of this is the differential extent to which rich and poor households use detergents for non-laundering purposes.

Table 7.6 shows that the two low-income groups use detergents for the washing up of dishes and cleaning of surfaces to a far greater extent than the high-income consumers who tend to use products specially formulated for these non-laundering purposes instead. The pattern of usage by the former, which reflects a common tendency for those in poverty to use a single product for a variety of purposes, is very probably

TABLE 7.6   The use of detergents for non-laundering purposes

	Urban poor	Rural poor	Urban medium	Urban rich
Of those using Breeze % using for dishes	88	90	58	23
Of those using Breeze % using for surfaces	92	93	85	48
Of those using Drive % using for dishes	84	74	61	0
Of those using Drive % using for surfaces	96	96	78	43

inefficient. Tests conducted by the major consumer group in the UK show, for example, that washing-up liquids are far more economical in washing dishes than detergents.[21] There is little reason to suppose that this result would not hold true in Barbados as well.

The 'traditional' orientation of the poor is manifest also in the way in which their laundering is conducted. Relatively few households in the low-income groups, for example, had read the instructions for the use of detergents and were thus less likely to use these products correctly (by, for example, pre-soaking the general wash).[22] The failure to read the instructions cannot entirely (or even largely) be ascribed to difficulties in understanding on the part of these consumers (almost all of whom are literate). There seemed, rather, to be almost a scornful attitude on the part of many to the use of instructions, reflecting perhaps a poor orientation to the need for precision in the use of many modern products. The problem may itself reflect the wider socioeconomic phenomenon of 'Barbados at the crossroads, caught up in the trans-ference from the ancient to the modern'.[23]

Nor, unlike the high-income consumers, did more than a tiny minority of the low-income groups wear gloves during the general wash. Since all detergents have at best some effect in removing natural grease from the skin and at worst produce an allergic reaction, those who do not wear gloves may be said to receive more of a negative characteristic than those who do.

In sum, then, some tendency for the low-income groups to derive fewer positive characteristics (or more negative characteristics) from synthetic detergents than the relatively affluent households can be discerned. It is our view that this tendency can be ascribed to the different way in which rich and poor consumers relate to the use of modern products.

The distributional impact of changing product characteristics in the industry

In reporting on the results of the case-study of Barbados we have thus far been concerned solely with its static component – with the choice, that is, from *given* laundry cleaning products. But as we argued in the theoretical analysis above, existing goods are themselves the result of past technical change which has added to the relevant part of the efficiency frontier for some individuals, left it unaltered for others and reduced it for yet others. In this section we attempt to determine the distributional impact of changed characteristics of laundering products.

To do so we shall need to trace the history of the major changes in the industry.[24]

The history of product changes

Laundry-soap production in Barbados dates back to 1943, during the war, when a Mr Roberts began production in a drum placed on top of a wood fire. By 1956 Roberts Manufacturing Company was producing 3.6 million lb per annum. The two Roberts products, one blue and the other brown, were sold in unwrapped form with the brand name 'Bomber' stamped on each bar.

In the early 1960s synthetic detergents were introduced in Barbados. These products possess a major advantage over soap in hard water in that they are not prone to calcium precipitation. The hardness of most of the water supplied in Barbados contributed to the rapid diffusion of synthetic detergents throughout the island. As Table 7.7 shows, this was accompanied by a steady decline in the production of laundry soap.

In November 1973 the Roberts plant, then the sole producer of laundry soaps in Barbados, closed down. The closure is an interesting example of how a declining product can be totally withdrawn from the market while a substantial residual demand for it still remains. The major reason was a variant of the process described above by which the old product is removed by the firm so as to divert demand to its newer version. In the Roberts case it was not the same firm that was responsible for both the old and the new products, but rather an entrepreneur with conflicting interests in both. Specifically, a major shareholder in the Roberts plant also owned a large importing agency. When a leading international firm manufacturing synthetic detergents appeared as possible clients for the agency, it made obvious business sense for the declining laundry-soap operations to be abandoned.

Did the demise of laundry-soap production affect all consumers equally? Clearly, whether and to what extent any particular household was adversely affected depends upon whether the laundry soap manufactured by Roberts was being consumed at the time and (if so) upon the perceived substitutability of alternative products. One may thus divide consumers at the end of 1973 into the following groups:

(a) those consuming only synthetic detergents
(b) those consuming only laundry soap(s) manufactured by Roberts
(c) those using synthetic detergents for the general wash and a brand of Roberts soap for ancillary purposes (such as stain removal)

TABLE 7.7   The decline of laundry-soap production

Year	Production of laundry soap (millions of lb)	Imports of synthetic detergents* (millions of lb)
1956	3.7	
1960	3.2	
1964	2.2	
1967	2.1	2.0
1970	2.7	2.5
1971	2.3	2.7
1972	2.1	3.2
1973	1.4	3.0
1974	–	4.1
1975	–	4.1
1976	–	3.6
1977	–	4.1
1978	–	4.6

* These estimates represent SITC category 554.2 which includes synthetic detergents for industrial as well as household purposes. The ratio between the two for 1978 was slightly greater than 1:1 in favour of household products. There seems little reason to suppose that this ratio differed much for the earlier years.

Dashes denote zero estimates. Blanks denote that figures cannot be obtained.

Sources: Barbados, *Abstract of Statistics* (1960); *Monthly Digest of Statistics* (Dec 1975); *Quarterly Digest of Statistics* (1969 and 1973); various issues of the *Annual of Overseas Trade*; various issues of the UN, *Yearbook of International Trade Statistics*.

(d) those using mainly a brand of Roberts laundry soap and occasionally a synthetic detergent. Only groups (b), (c) and (d) could therefore have been adversely affected. Whether in fact this was the case depends upon the substitutability of the alternative products available at the time as perceived by the members of these groups. The options available were to switch to another variety of laundry soap, to synthetic detergents, or to some other good (such as toilet soap). In order to establish the extent of these various possibilities, we need first to examine the consumption of laundry soap prior to the closure of the Roberts plant and in the period immediately thereafter.

The consumption of laundry soap manufactured by Roberts in 1973 was some 700 000 lb (i.e. production less exports). Who were its major users at the time? Although there is no direct evidence on this question one can none the less adduce some useful evidence of an indirect kind.

The survey showed, on the one hand, that while 4 or 5 per cent of the low-income groups consume only laundry soaps, this is not true of any high-income consumers. On the other hand, the proportion of respondents using only detergents is higher among the latter.[25] In 1973, with double the total present consumption of laundry soap, it is almost certain that these differences between low- and high-income consumers were more exaggerated, so that say 10 per cent of poor consumers used only laundry soap but that there were even then no rich individuals of whom this was true. Furthermore, evidence for neighbouring Trinidad and Tobago shows that low-income households spend a much higher percentage of total expenditure for laundry products on laundry soap than those with relatively high incomes.[26] It does not seem unreasonable to expect a similar pattern to prevail in Barbados.

For these reasons it would thus appear to have been mainly those with low incomes who could have been adversely affected by the closure of the laundry-soap plant. Whether or not this actually happened depends, as noted above, on the perceived closeness of substitutable products. Respondents in the survey were therefore asked whether they had had to switch to a less preferred product following the closure of the Roberts factory. The results are shown in Table 7.8.

That many of the low-income consumers felt adversely affected by the

TABLE 7.8   The perceived effect of the withdrawal of Roberts soap

	Urban poor	Rural poor	Urban medium	Urban rich
% who had to switch to less preferred laundry product from laundry soap produced by Roberts	39	51	23	10*
% who did not have to switch to less preferred laundry products from laundry soap produced by Roberts	44	33	66	59
% who had never used laundry soap produced by Roberts	17	16	11	30
	100	100	100	100

* One respondent who left the country shortly after the closure of Roberts was not required to answer this question.

withdrawal of the Roberts soap can be explained first by the nature of the substitute laundry soaps that were available at the time. One of the major alternatives to which consumers shifted after 1973 was Sunlight soap, a packaged variety manufactured by a large international company. This soap was some 40 per cent more expensive than the Roberts brands. However, it almost certainly embodied fewer cleaning characteristics per unit price. Another popular substitute at the time was an unpackaged brand known as 'Palm'. Although this product sold at the same price as the Roberts soap, a laboratory test that we commissioned showed (a sample of) the latter to be superior with respect to soil removal (but equivalent in stain-removing ability).[27] As with Sunlight, therefore, fewer cleaning characteristics per unit price were available to those who had to shift from the consumption of Roberts soap.

The fact that, as Table 7.8 shows, a far higher proportion of the high-income group had never used the latter is one reason why proportionately fewer high-income consumers felt adversely affected by its withdrawal. Secondly, we may suppose by extension that a greater proportion of rich consumers were not using the soaps immediately prior to their withdrawal. Finally, even among those who *were* using Roberts soap at that time, there is a reason to suppose that the proportion forced to switch to less preferred brands was higher in the low-income group than in the rich sample. This reason is based upon the differential use of the laundry soaps by the two groups. For many of those with low incomes the soaps manufactured by Roberts may very probably have been used largely for the general wash whereas in high-income homes their principal use was probably in stain removal. Since we have noted above that Roberts soap was found in the laboratory test to be superior to a major substitute ('Palm') with respect to soil but not stain removal, it is the low- rather than high-income users of the former who would thus tend to have suffered.

So far we have examined the distributional impact of the introduction of synthetic detergents indirectly via their effect on the substitute product, laundry soap. But we need also to examine the distributional impact of the synthetic detergents themselves.

There can be little doubt that 'the new synthetic detergent powders, so much more efficient at washing clothes in hard water, have helped women enormously'.[28] This does not mean, however, that some or all groups in Barbadian society could not have benefited even more from a product (or set of products) embodying different characteristics or the same characteristics in different proportions.

The vast majority of synthetic detergents available in Barbados embody developed country standards of functionality and packaging, and a number of them are also advertised on television and radio. Partly on account of these factors even the prices of products imported free of duty into the country are very high – higher in fact than synthetic detergents available in at least one developed country.[29] In terms of our argument in the first section of the chapter, the standards in developed countries tend to exceed those that would be chosen by the typical low-income consumer in the Third World. The low-income consumer, that is to say, would be prepared to trade-off developed country standards of quality for price reductions. The survey was designed to test this proposition in relation to one characteristic of synthetic detergents, namely, packaging. In particular, because the high-income consumers may be expected to value 'attractive packaging' more highly than the other income groups, the former groups should be less willing to trade this characteristic for a given price reduction than the latter.

Although 'attractive packaging' need not, of course, be synonymous with expensive packaging, we felt that in the context attractiveness was indeed associated in the minds of most consumers with coloured, expensive, cardboard packages. It was therefore decided to ask re-spondents to conduct a thought experiment in which the familiar medium-sized box of Breeze packaged in brightly coloured cardboard was compared with a notional packet of Breeze wrapped in plastic. The latter was to be considered equivalent to the regular packet of Breeze in all respects save only for the packaging. Respondents were then asked how much (if any) of a reduction in price below that of 199 cents (the price of the medium-sized box of Breeze) would be required before the notional product packaged in plastic was chosen in preference to the Breeze in the regular packaging. The results of the thought experiment are shown in Table 7.9.

Evidently, a very much higher proportion of the high-income group would require a price reduction in excess of 9 cents (5 per cent) before they would prefer to buy Breeze packaged in plastic rather than cardboard. Some support is thus provided for our view of income as a fundamental determinant of preferences (defined in terms of charac-teristics). Similar results to those obtained in the case of packaging may also be expected for the other characteristics of detergents that conform to developed country standards, particularly those such as smoothness of texture, soil- and stain-removing ability, etc. (though it is difficult to devise a thought experiment for these characteristics that is comparable to that described above for packaging).

TABLE 7.9 The 'trade-off' of attractive packaging for a given price reduction*

	Urban poor	Rural poor	Urban medium	Urban rich
% of respondents requiring a reduction of less than 9 cents before notional Breeze is preferred to actual Breeze	71	77	47	22
% of respondents requiring a reduction of more than 9 cents before notional Breeze is preferred to actual Breeze	29	23	53	78
	100	100	100	100

* 21 per cent, 8 per cent and 1 per cent of the responses in the urban poor, rural poor and urban medium groups respectively were rejected as the interviewer had established the price reduction needed to leave the consumer in a state of indifference rather than preference.

Before one concludes from this discussion that the characteristics of detergents in Barbados favour the high- rather than the low-income consumers, it has, however, to be pointed out that less expensive brands without elaborate packaging or heavy promotional expenditures have been available in varying quantities since the mid-1960s. These products are mostly imported in bulk and packaged locally in simple plastic bags. At a price some 30 per cent below the cheapest of the detergents packaged in cardboard it is not surprising that these brands have been popular with the low-income groups.

Undoubtedly, the availability of the detergents packaged in plastic has reduced the inegalitarian distributional impact associated with the production of 'high-income' products. For several reasons, however, this offsetting influence has been only partial. First, the price reductions have, on occasion, been achieved at the expense of unacceptably low quality. In one extreme example of this point, many consumers burnt their hands as a direct result of the use of a detergent wrapped in plastic which had been imported from Canada and was found to possess excess alkalinity. The second reason is that the inexpensive varieties, unlike the leading brands, have not been generally available in the small retail outlets and especially those in rural areas.

The bulk of the distribution of the plastic-packaged products is undertaken by a single individual who stressed the high opportunity costs of having to sell very small quantities to retailers in isolated rural localities. Since it is the small outlets from which proportionately more low- as opposed to high-income consumers generally purchase detergents, it is thus the former individuals who have less opportunity to exchange 'high-income' characteristics for a lower price.

7.3 Conclusions and Policy Implications

In this chapter we have been concerned to analyse two basic issues; first, the static question of the efficiency of the consumption process and, secondly, the dynamics of changed product characteristics and the impact of these changes on the distribution of income. Both problems were analysed in the context of the consumption of laundry soaps and synthetic detergents in Barbados.

As far as the static question is concerned, we found all income groups to be poorly informed about the cleaning characteristics of several major brands and some tendency for greater inefficiency in the other components (divisibility, accounting and product use skills) of the consumption process among the low-income consumers. Similarly, in examining the distributional impact of changed product characteristics in the industry it is the poor who have tended to lose relatively and, to some extent, in absolute terms as well.

These findings suggest that there may be a number of ways in which the lot of the poorest consumers in Barbados can be improved *without* a redistribution of income in their favour. In particular, we advocate that the following complementary policy measures be adopted.[30]

Education and consumer skills

In section 7.2 of this chapter it was argued that inefficiencies in the consumption process of low-income consumers cannot be solely attributed to problems of information and computation, although of course these factors are important. It is our view that the causes of many of the inefficiencies are more fundamental – based on the attitudes of the poor to the consumption process in general. If these attitudes are to be changed the process will almost certainly need to begin at the formative level of the education system. At present this system pays no attention to consumer education; on the contrary, secondary-school teaching in the Caribbean is 'academic rather than practical, focused on Europe rather than on the Caribbean, affording prestige rather than social utility'.[31]

It is difficult to specify in detail the content of consumer education programmes for the different age groups of school children. However, since the majority of children from the poorest households in Barbados do not reach secondary-school level, it is important that this type of education be introduced in both primary and secondary schools. Moreover, consumer education in the broad sense could usefully be integrated into the operations of the mass media, and into the adult education system.

Consumer organisations

The record of consumer organisations in Barbados is a dismal one. Thus, 'over the past decade, several consumer bodies have been formed, but lack of numerical support and sometimes internal manoeuvring for power or prestige, have resulted in their demise'.[32]

One important reason for the lack of support for these organisations is that most Barbadians (especially the low-income groups) are not highly oriented to those needs – for careful shopping, protection from producer malpractices, etc. – on which much of the appeal of consumer organisation depends. It is unfortunately true, therefore, that 'Barbados will need to be educated about consumerism and take the information to heart . . . before any new consumer body can hope to succeed'.[33]

Even without consumer education – which as we suggested above should be introduced into the education system – there is none the less a good deal that even a small group can achieve by public exposure of some of the more serious manifestations of the highly inequitable relationship between producers and consumers that currently exists.

The formulation of standards

Though a Standards Institution does exist in Barbados its role should be greatly extended. Moreover, unless the activities of the Institution are co-ordinated with other policies to improve the efficiency of the consumption process, few of the benefits of its work will be felt by consumers. Though the Barbados Standards Institution has, for example, undertaken comparative testing of the various chlorine bleaches on the market the results have not been conveyed to the consuming public.

There is no reason why the standards laid down for detergents (or any other product) should conform to those adopted by the developed countries. The effect of this would be – as we saw in the case of expensive packaging – to penalise those with low incomes. But it would be very helpful to consumers to know that the detergents available on the market conform to *some* standards of quality (albeit *different* standards). In this respect a simple grading system operated by the Standards Institution could be very valuable (especially in reducing the risk involved in purchasing cheaper, less well-known brands).

The provision of information

As suggested above, the work of an extended Standards Institution needs to be communicated to consumers. Because radios and (to a lesser extent) televisions are widely owned, these media are probably the most effective means of communicating information about product characteristics. It is important that the programmes offered are lively and entertaining so that the widest possible audience is reached. To be effective, information should be of a *comparative* kind – there is little point in merely conveying to consumers the characteristics embodied in a large number of different brands of detergents.

In addition to the provision of comparative information, radio and television consumer advice programmes should offer guidance on the correct use and combination of products, the advantages of unit price calculations and the 'filtering' of advertising claims.

Local production of synthetic detergents

Thus far our policy proposals relate mainly to the need for improvements in the efficiency with which given laundry cleaning products are consumed. But there is also a substantial case for a policy which, by extending the goods available and altering the terms on which existing goods can be purchased, increases the welfare of consumers and especially those with relatively low incomes. What is required in particular is government encouragement of 'appropriate' products – that is, synthetic detergents which allow basic cleaning needs to be cheaply met, so counteracting the monopolistic pricing practices of international synthetic detergent manufacturers and creating much needed direct and indirect employment on the island.

Notes

1. At the time this work was being completed the author was Research Fellow at Queen Elizabeth House, Oxford; he is presently working with the ILO, Geneva.

2. T. Scitovsky, *The Joyless Economy: An Inquiry into Human Satisfaction and Consumer Dissatisfaction* (Oxford University Press, 1976) pp. 4–5.
3. K. Lancaster, 'A New Approach to Consumer Theory', *Journal of Political Economy*. vol. lxxiv (Apr 1966).
4. These are summarised in M. D. Vernon, *The Psychology of Perception* (Harmondsworth: Penguin, 1962).
5. Ibid, p. 184.
6. Ibid.
7. R. W. Olshavsky and J. A. Miller, 'Consumer Expectations, Product Performance and Perceived Product Quality', *Journal of Marketing Research* (Feb 1972).
8. D. A. L. Auld, 'Imperfect Knowledge and the New Theory of Demand', *Journal of Political Economy*, vol. lxxx (Nov–Dec 1972).
9. Formally, the requirement of perfect divisibility is that if $x = (a, b)$ and $x' = (a', b')$ are any two attainable vectors of quantities of characteristics, then any vector $kx + (l-k)x'$ also be attainable (with $o \leqslant k \leqslant l$). See H. A. John Green, *Consumer Theory* (Harmondsworth: Penguin, 1971).
10. See J. James and F. Stewart, 'New Products: A Discussion of the Welfare Effects of the Introduction of New Products in Developing Countries', *Oxford Economic Papers*, vol. 33 (Mar 1981).
11. Following Hans Brems, 'Allocation and Distribution Theory: Technological Innovation and Progress – Discussion', *American Economic Review*, papers and proceedings, vol. 56 (1966).
12. Victor Papanek, *Design for the Real World* (London: Paladin, 1974) p. 156.
13. F. Stewart, *Technology and Underdevelopment* (London: Macmillan, 1977).
14. The following analysis is based on James and Stewart, 'New Products: A Discussion of the Welfare Effects of the Introduction of New Products in Developing Countries'.
15. For a full description of the survey design and the respondent profile see the first appendix in Jeffrey James, *Product Choice and Poverty: A Study of the Inefficiency of Low-income Consumption and the Distributional Impact of Product Changes* (Geneva: ILO, 1980; mimeographed World Employment Programme research working paper; restricted).
16. For an excellent introduction see J. B. Kruskal and M. Wish, *Multidimensional Scaling* (London: Sage, 1978). Other references are to be found in James, *Product Choice and Poverty*, p. 46.
17. See, for example, P. E. Green and V. R. Rao, *Applied Multidimensional Scaling: A Comparison of Approach and Algorithms* (New York: Holt, Rinehart & Winston, 1972). See also the discussion and references in James, *Product Choice and Poverty*, pp. 48–52 and Appendix 5.
18. The high-income group take a generally less favourable view of the laundry soap and a more positive view of Tide compared to the other three groups. See James, *Product Choice and Poverty*, pp. 72–81.
19. For a description of the test, see ibid, p. 64.
20. The annual expenditures for advertising of brands available in Barbados can be found in ibid, p. 31 (see also pp. 32–6).
21. *Which?* (Oct 1965). See also James, *Product Choice and Poverty*, p. 42.
22. Table 9 (p. 43) in James, *Product Choice and Poverty*, summarises the different washing practices of the four income groups.

23. G. S. Dann (ed.), *Everyday Life in Barbados: A Sociological Perspective* (Leiden: Department of Caribbean Studies of the Royal Institute of Linguistics and Anthropology, 1976) p. 35.
24. A fuller description is to be found in section 3B of James, *Product Choice and Poverty*.
25. Specifically, 66 per cent of the high-income group use no laundry soap as opposed to the figures of 47 per cent and 51 per cent for the urban and rural poor groups respectively.
26. Central Statistical Office, *Household Budgetary Survey* (Trinidad and Tobago, 1975–6).
27. See James, *Product Choice and Poverty*, Appendix 4.
28. *Which?* (Sept 1959) p. 115.
29. Towards the end of 1979, for example, a 29.6-ounce packet of Drive sold in England could be purchased for 54 pence. In Barbados the product of the same brand name cost 390 cents for a 26-ounce package. Thus, the price per ounce in Barbados was nearly twice that in England.
30. A fuller discussion of these proposals is contained in James, *Product Choice and Poverty*, pp. 123–33.
31. David Lowenthal, *West Indian Societies* (Oxford University Press, 1972) p. 120.
32. Barbados, *Sunday Advocate News*, 22 July 1979.
33. Ibid.

8 Weaning Food and Low-income Consumers in Ethiopia

Marie-Ann Landgren-Gudina[1]

Weaning food is the basic food for children aged between six months and two years, that is, in the period between full breast-feeding and adult food. Basically, one can distinguish four types of weaning food in Ethiopia:

(a) traditional weaning food which is a by-product of the food eaten by adults
(b) infant formula/milk which is either imported (and/or donated) milk powder or cow's milk
(c) industrial weaning food which is either produced by the only local producer (Faffa Food Plants) or imported from abroad
(d) home-made weaning food prepared according to prescriptions of the Ethiopian Nutrition Institute (ENI).

The main objective of this chapter is to determine which of these products is most appropriate for low-income classes. To that end a small consumer survey was carried out in some 200 urban and rural households in western Ethiopia. Another objective is to examine the employment consequences of alternative technologies in producing weaning food. Since there is only one local industrial producer and most weaning food is prepared at home, this aspect will receive less attention in this chapter.

Chapter 8 is organised as follows. Section 8.1 provides some background information on Ethiopia, and on some of the social changes that have taken place since 1975. Sections 8.2 and 8.3 examine the role of the Ethiopian Nutrition Institute (ENI) and the Faffa Food Plants in the preparation and production of weaning food. Section 8.4 describes the survey area and gives some information on the survey itself. In the fifth section (8.5) the results of the consumer survey are presented, revealing the extent to which various income groups purchase the four types of weaning food available in Ethiopia. In addition, it examines the incidence of breast-feeding, bottle-feeding and different types of weaning food in the different income groups. Finally, it discusses some desirable characteristics of weaning food. The concluding section (8.6) gives an estimate of future demand for weaning food and provides some recommendations as to how to fulfil this demand.

8.1 Ethiopia – The Country of Investigation

Ethiopia is one of the poorest nations in the world. GNP per head was about US$100 in 1976 and the annual growth of GNP was not more than 0.2 per cent between 1970 and 1976. In 1970, 84 per cent of the population was occupied in agriculture, and only a small fraction in mining and industry (6 per cent) and in services (10 per cent). In 1973 the shares of these sectors in total GNP were: in agriculture, 55 per cent; mining and industry, 16 per cent; and services 29 per cent. It is unlikely that these proportions have changed significantly.[2] Most farmers live at the subsistence level in rather isolated villages, far from the few roads. Illiteracy is very high, perhaps around 90 per cent in many rural areas.[3]

The landholding system was feudal[4] until the Provisional Military Administrative Council (PMAC), better known as the Dergue, made two important proclamations in 1975: the Proclamation on Rural Land Reform[5] and the Proclamation on Urban Land and Extra Houses.[6] Both proclamations are important for this study as surveys were carried out both in rural areas and in towns. The Proclamation on Rural Land Reform provides for all land to be given to the tiller, and forbids anyone to farm an area of more than 10 hectares. In addition, all farming households, with an average of 200 families on 800 hectares, are organised in co-operatives which are called peasant associations. The co-operative is responsible for land without an owner (confiscated land) and decides on what to do with it. Each member of the co-operative farms spends a fixed amount of time a week on the communal farm. Farmers must pay land tax to the government and the collection of this tax is also the responsibility of the co-operative.[7] The latter is also responsible for social welfare but in most cases this function is unfulfilled even four years after its inception.[8] Within the peasant association there are different groups, such as the youth group and the women's group. Most decisions concerning the area are taken in the co-operative: for example, who will be sent to the militia (a type of armed force) and who will be given a poverty certificate in order to receive free medical help and free school material. The farmers elect members to a higher-level co-operative called a service co-operative, the latter having a mediating function between the *woreda*, or local government, and the co-operative.[9] The most important task in the long run is the reallocation of land according to the needs of the different members, without allowing anybody to have more than 10 hectares. In most peasant associations this work has not started as there are many other acute problems.[10]

In the Proclamation on Urban Land and Extra Houses[11] it is stated that each person is allowed to own or to build one house to live in. If the owner does not live in his own house, it is nationalised. An exemption is made for children who are too young to have a household of their own. Richer families took the chance to distribute their houses among their children. These houses may be rented until the children are old enough to have a house of their own.

At the same time, the house rent was reduced according to a special scale. Those who live in town were put together into urban dwellers associations or *kebeles*. The number of *kebeles* varies with the size of the town; there are six in Ghimbi, the town to be surveyed in this study, and nineteen in Addis Ababa. The *kebele* is responsible for the area; it registers all those who live in it; it manages the special co-operative shops (where one can buy scarce products on presentation of a registration card); it is also responsible for law and justice and the prisons. As with the peasant associations, the members (those who live in the area) elect representatives at the local and higher levels. As in rural areas, the *kebele* can give a poverty certificate which gives the holder the right to free medical care and so on. In order to obtain financial means for the *kebele*, house rents up to a special monthly rate are collected by the *kebele*. Certain houses with higher house rents pay their rents directly to the Ministry of Housing.

8.2 The Ethiopian Nutrition Institute (ENI)

The Ethiopian Nutrition Institute was the first institution in Ethiopia to be concerned with weaning food. Until 1968, when it was called the Children's Nutrition Unit, it was mainly engaged in surveys, often in collaboration with children's clinics. At the same time it developed the formula for 'Faffa' (see Section 8.3) which is a low-cost weaning food. It also started to work on home-made weaning food. Their main guides were King[12] whose book deals mainly with cassava- and maize-eating cultures and Cameron and Hofvander[13] whose book is more general and gives recipes based on staple grains. This last book is widely used in Ethiopia, both at the ENI and in the various clinics that give nutrition education to mothers and pregnant women.

In the 1970s the ENI proposed a new infant food which was then produced by Faffa. In addition, the Institute diversified its activities into nutrition education for the whole family, training for nutrition extension agents, courses on home economics, and programmes with peasant associations (in particular on gardening). It has also renewed its efforts on weaning-food recipes. The big problem is that the food is too bulky, with the result that the child is already satisfied before it has taken sufficient nutrients. Another problem is that weaning food should be a by-product of the food normally consumed by the family. (In most homes there is only one open fire and it is almost impossible to prepare more than one type of food at once.)

8.3 Faffa Food Plants

In 1967–8 the ENI started to produce Faffa, a low-cost weaning food. Faffa is a powder from which a porridge or gruel can be prepared. One reason for choosing porridge is that it is a well-known traditional food in Ethiopia; another is that it can be diluted with water; and a third is that it is within the means of low-income earners.

Production

In 1976 the Faffa Food Plant became independent of ENI, and in 1979 a new factory was opened. The information on the new factory gathered for this study was collected in September/October 1979, only a few months after the opening in April. As a result, the factory was producing at about 72.5 per cent capacity: some of the machinery was still defective, the workers had still to familiarise themselves with the new machines and there were occasional shortages of water, electricity and fuel. The plant has a capacity of 12 000 metric tons per year[14] and employs some 151 persons. Table 8.1 gives a percentage distribution of its products in 1978.

All the products in Table 8.1 were developed in Ethiopia and are not produced under any licence. The production figures on which the table is

TABLE 8.1 Products manufactured by Faffa Food
Plants (1978)

Product	Brand	% of output
Weaning food	Faffa	51
	Meten	3
Baby (infant) formula	Edget	6
Fortified wheat flour	Dube Duket	40
Total		100

Source. Faffa Food Plant Report.

based were measured in terms of metric tons and not in terms of value,
because the products are sold at different prices in different markets.
Faffa Food Plants have always produced at a loss, in spite of the fact
that they are exempt from customs, transaction, turnover and profit
taxes. Other companies must pay 12 per cent on raw materials, 7 per cent
transaction tax and 2 per cent turnover tax. The Swedish International
Development Association (SIDA) is subsidising Faffa, in some years at
the rate of 60 per cent of production costs.

Almost all the ingredients of the weaning-food product 'Faffa' are
manufactured in the plant. These ingredients are: wheat grains,
field/chick peas, full-fat soybeans, sugar, salt (iodised) and wheat flour.
On the other hand, many of the ingredients for the newer product
'Meten' are imported. (These are: full-fat milk, non-fat soya flour, non-
fat milk, calcium phosphate, ferrum reductum (iron) and vitamin
premix.)

The production processes include milling (mainly sugar), roasting,
extruding, mixing and packing. Most machinery is of European origin
and recently purchased with a SIDA grant.

Marketing

All weaning food produced by Faffa is sold within Ethiopia. Up to 1978,
50 per cent went to the Relief and Rehabilitation Commission, 40 per
cent to retailers and 10 per cent to institutions such as hospitals and
clinics.

The poorest consumers are unable to buy a packet of Faffa, even at
the subsidised price of 25 Ethiopian cents (in 1979 1US$ was 2.05 Birr,
i.e. 100 Eth. cents) sufficient only for three days. ENI considers that only
5–15 per cent of the population is within reach of a product like Faffa on

the commercial market, either economically or geographically. The target groups are the middle-income groups in town, who adopt innovations like weaning food more quickly than the rural population. High-income earners can afford to buy imported infant formulas and weaning food but they often prefer to buy Faffa because they know that it is cheap and nutritious.[15] Today the high-income groups are the target group for Faffa Food Plant's new products, such as Meten, Edget and Dube Duket.

Faffa's commercial network is thinly spread over the country. It includes five sales regions each covered by one salesman, except in Addis Ababa where there are two. Each salesman has a delivery van and a helper and some minor changes have been made in order to take advantage of the Ethiopian Government Transport Organisation (Ketena). It is envisaged to restructure the marketing organisation so as to take advantage of the peasant associations and *kebeles*.[16]

There is a shortage of Faffa on the commercial market because the Relief and Rehabilitation Commission (RRC) takes so much for relief purposes. In 1976–7 more than a third, and in 1977–8 almost two-thirds, and in 1978–9 more than 85 per cent was so consumed. The consumers who are able to pay must either buy it or obtain it freely from the institutional market.

Faffa was designed for households with an income of about 50 Birr a month when it was first introduced in 1968. Today it is consumed by households with incomes below 50 Birr a month who obtain it via relief organisations or institutional markets as most victims are in this income group. Most of the food available in the open market is purchased by households with incomes in the range of 50–100 Birr.

Today the Faffa Food Plant produces products other than Faffa. This idea came up first in the *Plan for Diversification*.[17] It was clear that the Ethiopian market needed a product that could be consumed by babies below the age of four months, as some mothers cannot breast-feed their children. Instead of buying imported formulas or milk powder they started to produce Baby Faffa, today called Edget. In 1976–7 the rate of production was below 14 tons per annum and in 1978–9 it increased to more than 270 tons. Meten is a new product which is a competitor to Faffa on the commercial market. The difference is that Meten is pre-cooked and it can be mixed with warm water and given instantly. In fact Meten is much cheaper than imported weaning foods: 400 grams cost about 5 cents in 1979 while 400 grams of Cerelac cost about 10 cents and 300 grams of Farox about 18 cents. Meten and Edget are not subsidised, but are planned to sell at a profit, the surplus to be transferred to Faffa in order to reduce the loss. In addition to these products the plant is also producing a soy-wheat flour called Dube Duket.

For the future the Faffa Food Plants are working with a soybean project in Wollega. They are also trying to develop Faffa from maize and Meten from local soybeans. There is a shortage of raw material in the country and the factory is forced to have a stock for nine months ahead.

8.4 Description of the Survey Area and the Survey

Given the limited resources available for the consumer survey, it was necessary to select a region that included both urban and rural areas, developed and less developed. For this purpose the Ghimbi *awraja* was chosen because it had two further advantages: first it is accessible by road, and secondly considerable data on the region were already available. Finally, the majority of the population belongs to one tribe, the Oromos.[18]

The Ghimbi area is part of the Wollega province in west Ethiopia. This region is a tropical highland region 1500–2000 metres above sea level, and is intersected by deep valleys. The Blue Nile is the borderline in the north. The Oromos were originally pastoralists, but when rinderpest came to the area some ninety years ago they had to change their way of life and become farmers.[19] The ox-drawn plough and *tef*, the cereal unique to Ethiopia, as well as maize, came to this area from the agriculturally advanced areas north of the Blue Nile. Farming in the highlands is predominantly subsistence farming and today the main crops are *tef*, maize, sorghum, finger-millet, supplemented by various legumes, pulses and root crops; the cash crop is coffee. In the past farming yielded rich harvests. However, as in other parts of Ethiopia, soil erosion is a problem resulting from the destruction of forests, intensive grazing and cultivation.

Food habits in western Wollega

The most common food in western Wollega, as in most of the Central Ethiopian Highlands, is *enjera* and *wot*. Other food, such as maize, yams and Galla potato, are also eaten.

Enjera is a thin pancake-like, sour, leavened bread which can be made out of any cereal available, usually *tef*. Since *tef* was rather expensive at the time of the survey, other cereals such as maize, sorghum and barley were being used. The weight of an *enjera* depends on the size of the *mitad* (baking pan) and the cereal used. *Tef enjera* is normally 350–450 grams and one of maize between 400–500 grams.

Enjera is normally eaten together with *wot*. Its main ingredients are legumes, meat chicken, vegetables or tubers. Onion, fat (butter or oil), salt and spices are added. The Ethiopians prefer to eat the *wot* with large quantities of fat (oil during fasting days for Ethiopian Orthodox). Wealthy people prepare their *wot* with large amounts of protein-rich food (meat, chicken or legumes), whereas poor people will have more watery *wot* with less or no fat and small quantities of protein-rich food. These families will also mainly serve dishes prepared with legumes, vegetables or tubers, as they cannot afford to buy meat or chicken.[20]

In western Wollega *enjera* is often eaten together with yams prepared as *wot* and the Galla potato is also consumed frequently. Towards the end of the rainy season (September onwards) and until the harvesting of cereal crops starts (December–January) maize is the main food consumed by the poorer families. If the family can afford it, they give one *enjera* to each adult. All the *enjeras* for one meal are put on one large plate. The *wot* is served in a bowl and spread over the *enjera*, so that it will be soaked before the meal starts. Then the father and the guests (if any) start to eat, sometimes together with the older boys. They eat most of the protein-rich food. Afterwards the mother and the children eat the leftovers. The thin part of the *wot* and the soaked *enjera* is sometimes the only food left. Small children eat slowly and cannot protest, with the result that they are often left only with *enjera*.

Fitfit is a mixture of *enjera* and sauce. It can be served to the whole family, but is especially prepared when food must be taken to other places or to the field. The mixture is often given to a child when it begins to follow the family diet. Fresh milk is available in limited quantities, because the daily production of a cow is often not more than 1 litre. It is kept especially for infants, but the main part is used to make butter. The buttermilk can be used as a beverage or may be eaten with *enjera* for breakfast, or to make a low-fat sourmild cheese (*ayib*). *Ayib* has a high protein content (about 15 per cent of its total weight).[21] Butter, unspiced, is also given to small children. It is put into their mouths and noses to grease the intestines. The butter is also used by the Oromo women for greasing their bodies and the hair. Most butter is made into ghee. The ghee is sometimes sold to get the money needed to buy salt and other goods.

The children of poor families are sometimes given the *enjera* which is used as a filter for producing local beer or honey-wine. Tea is sometimes served in richer families. Most families do not prepare special food for the children, but say that they eat with the family. In this part of the country they do not force a child to eat or to taste anything while it is breast-feeding; they simply wait until the child is old enough to stretch its hand and wants to taste the family food. At this moment the child will be given a piece of *enjera* or maize. The child will be given *enjera* or maize without sauce until it shows an interest in the sauce, when *fitfit* is given. The family never buys anything special for the children and they must eat with the family. During this period the mother continues to breast-feed the child and the protein in the breast-milk is very important.

The survey

Within the Ghimbi region four areas were chosen; Ghimbi Town, Bodji, Aira and Worra Djirru. This provides a reasonable sample of the different parts of the region. In each area it was decided to interview about fifty households together with the children. Questionnaires were prepared in English and Amharic, although the enumerators were given enough time to prepare their questions in Oromo (the local language), if respondents did not know Amharic. The interviews took place in October–November 1979, between ploughing and harvesting, a time of year known as 'the hungry period'.

In Bodji and Aira, which are extensive areas, people were interviewed in official places, on market days and on visits to the clinic. In Worra Djirru one area in the village was chosen and all the houses in that area were canvassed. In Ghimbi Town two *kebeles* were contacted and fifty houses were selected randomly.

One of the questions in the survey concerned income. Since it was mainly the women who were interviewed, it was not always possible to obtain correct information on either income or expenditure. Initially, it was planned to use the scale for taxation of agricultural income which distinguishes four income brackets: 1–600 Birr, 601–900 Birr, 901–1200 Birr, and above 1200 Birr per annum. Since most people reported income in the lowest brackets, they were asked to state the precise amount of their income which was later used to construct the income distribution. Figures on farmers' incomes are not complete since they do not include subsistence income.

8.5 The Survey Results

The time of year when the survey took place (October–November 1979) is a hard one for farmers, falling between ploughing and harvesting. However, if the maize is starting to ripen the people tend to eat a lot of it (sometimes in *enjera*).

At this season, there is quite a difference in nutritional status between rural and urban areas. On the one hand, in Worra Djirru everybody has a plot of land and each relies on own production for food consumption. In contrast, in Ghimbi Town everyone has to buy their food items. Many children in this town have marasmus because the family cannot afford enough food.

Another factor explaining the incidence of malnutrition, bottle-feeding and the use of home-made weaning food is the level of education. The level of education in Worra Djirru, for example, is lower than in Aira and Bodji, where some people had spent more than six years at school. Finally, it should be noted that Worra Djirru has no medical or nutritional facilities.

Breast-feeding

In the literature on small children most authors assume that the mothers in rural areas breast-feed their children longer than in urban areas. At the same time it is said that breast-feeding is more common among low-income groups. Both statements are confirmed by the survey, as Table 8.2 indicates.

None the less, there is quite some variation even in rural areas. In Worra Djirru, for example, no mother breast-fed her child for longer than two years. When these mothers were asked why, the standard reply was that the children's teeth would be destroyed if the breast-feeding went on beyond two years.

TABLE 8.2   Breastfeeding (in months) by income group (in percentages)

| Income classes | Three rural areas | | | | | Ghimbi town | | | | |
	1–12	13–18 (months)	18+	Not stated	Total	1–12	13–18 (months)	18+	Not stated	Total
I	10	30	60	–	100	36	18	45	–	100
II	9	24	67	–	100	19	52	29	–	100
III	9	22	67	2	100	11	67	22	–	100
IV	18	9	59	14	100	10	60	30	–	100
V	38	–	54	8	100	36	36	18	9	100
Total	13	20	64	3	100	23	47	29	2	100

Casual observation in other parts of Ethiopia shows that almost all women breast-feed their children. Most parents cannot afford regularly to buy the required quantity of infant formula. In addition, the hygienic conditions surrounding the preparation of the baby-food imply a health risk. In the larger towns, especially Addis Ababa, breast-feeding has decreased, a tendency now also spreading to smaller towns and rural areas.

Bottle-feeding

The practice of bottle-feeding is spreading all over the world and Ethiopia is no exception. Commercial radio programmes and publicity stress the importance of the bottle and many mothers use it while breast-feeding at the same time. In many cases mothers start the bottle when the time for weaning food (around six months) has arrived.

In Ethiopia, most women do not use the bottle. In Ghimbi Town, about 54 per cent used the bottle but in the three rural areas the figure was only 22 per cent. In rural areas, the bottle is mainly used to feed the child with cow's milk and in some cases water or tea. If the mother is not able to supply milk a UNICEF programme provides such mothers with milk powder. The main reason for using the bottle is that the mother is working in the field so that one of the older children can feed the child in her absence. In more remote areas such as Worra Djirru the bottle is hardly used, since one cannot buy bottles in the village. In urban areas, such as Ghimbi Town, mothers of middle- and high-income groups can afford to give their children infant formulas. It is exactly this group that is the target group for Faffa and other industrial-made weaning foods.

Different types of weaning food

In this study four weaning foods are distinguished: first, the traditional weaning food which is similar to foods eaten by adults such as *enjera* and *wot* (see section 8.4); secondly, imported milk powder or infant formulae and the locally produced infant formula Edget; thirdly, the industrial-made local weaning formulae, such as Faffa and Metan; and fourthly, the home-made weaning food, prepared according to ENI recipes (see Table 8.3).

It is interesting to note here that the infant formulae tend to be consumed in Ghimbi Town but not so much in the three neighbouring rural areas, but this is not true of the manufactured weaning food. Moreover, rural households tend to consume more home-made weaning

TABLE 8.3   Type of weaning food used by rural and urban households, by income classes (in %)

Type of weaning food	Three rural areas						Ghimbi Town					
	Income class						Income class					
	I	II	III	IV	V	Total	I	II	III	IV	V	Total
Traditional	75	67	61	73	50	64	91	58	45	55	35	54
Infant formula	8	9	10	9	6	9	–	29	36	18	40	28
Industrial	8	8	8	–	11	7	9	10	5	–	10	7
Home-made	8	16	22	18	33	20	–	3	14	27	15	11
Total	100	100	100	100	100	100	100	100	100	100	100	100

food than urban households and high-income groups (presumably more literate) are more likely to use this type of weaning food.

In western Wollega there is a shortage of industrial weaning-food products, although there is an all-weather road. No shops in Bila had Faffa or any other weaning-food products; only the clinic had them in stock. Aira is situated on a dry-weather road, and only the hospital could sell weaning food. Worra Djirru had no shop with weaning food, even though the all-weather road passes through it. The nearest place to buy weaning food is in Nedjo, three hours' walk from Worra Djirru.

In an urban area, Ghimbi, the situation was different. Most shops or kiosks had at least one brand on sale. One could find Cerelac 4:75, Morinaga 5:75, Sobee 5:25, S26 5:75 and Meiji 6:25. (All prices are for one can.) In no shop could one find Faffa and only one seller could offer Edget and Meten, both at the recommended price. Many families wanted to buy Faffa, but it was not available. Only those with higher incomes could afford to buy the more expensive brands available.

The clinic in Bodji distributes Faffa and milk powder freely to families whose children have too low a weight. In Aira, very needy children can receive Faffa and milk powder freely, but they are more strict in the distribution than the Bodji clinic. In Worra Djirru there is no clinic and no programme is found in the neighbourhood concerning weaning food. Ghimbi has one governmental clinic and one hospital and both give information about Faffa but not about the home-made alternatives. There is no free distribution of Faffa or milk powder in Ghimbi.

The majority of families give traditional food to their children. They normally begin to give *enjera* to the child when it is six or eight months old. This is too late according to experts. The recommended feeding plan for a developing country is as shown in Table 8.4.

TABLE 8.4 Recommended feeding plan for a developing country

	0	3	6	9	12	15	18 months
Breast milk							
Soft fruits or juice							
Double mixes							
Multimixes, other fruits							
Adult food							

Transitional period ⎯ ⎯ ⎯ ⎯
Time when food is regularly given ⎯⎯⎯⎯

Source. Cameron and Hofvander, *Manual on Feeding Infants and Young Children* (United Nations, 1976) Figure 29, p. 121.

The use of home-made weaning food is the most interesting phenomenon in this region. No peasant association had tried to produce local mixtures and consequently no data could be collected. Instead the hospital and clinics in rural areas had started a programme to teach mothers (including pregnant women) to prepare home-made weaning food from ingredients already at hand. ENI is producing material and has printed recipes both in English and Amharic. Besides this, the Cameron and Hofvander Manual[22] has one chapter with different mixes. The clinics and hospitals collect local data on food and try to adjust the given recipes for local use. Both in Bodji and Aira there are home economists employed to take care of the demonstrations and also the medical personnel are involved in the programme. It is very important that the mothers are allowed to taste the food, because no mother will prepare food for her child that she does not like herself.

Sometimes the staff give simple advice like: 'Do you have hens?'; if the answer is yes, staff say: 'then you can give your child an egg'. Such advice is relevant in Aira where the importance of eggs is stressed because eggs are available. Meat is more scarce and as it is rather a dry area it is hard to grow vegetables. In Aira the Nutrition and Rehabilitation Centre tries to teach people how to grow carrots, tomatoes and other vegetables and advice is also given about how to use them in the daily diet. The next step is to go out to the peasant associations and teach all the farmers how to grow vegetables and how to feed small children with this type of food.

Desirable characteristics of weaning food

In the survey there was also a question about the desirable characteristics of weaning food. A requirement mentioned by most respondents was that it ought to be as close to the normal family diet as possible. Porridge or products that looked like milk were appreciated most. Very few respondents would give the children a product like gruel because it is not a common food, especially among poor families. Another requirement frequently mentioned was that the taste should be acceptable to adults also; sweet products are highly appreciated. Liquid food that can be given by bottle is the most easily accepted, porridge comes second, but snacks or baby biscuits (as a meal) were not so acceptable. Several respondents placed great faith in the package and above all valued well-rounded babies on the cover's picture.

8.6 Conclusions

Probably the best way to improve the quality of weaning food in rural areas is to work through the peasant associations or co-operatives. Although they did not work efficiently immediately after the revolution in 1975, they have now gained more organisational experience and could be deployed in nutrition programmes. Naturally, the ENI should advise them on the content of these programmes, but they have too few staff to carry out programmes all over Ethiopia.

Peasant organisations can contact somebody on a higher level (for example, the nutrition agent) and ask for advice. They can be given recipes and ideas on how to prepare weaning food on the basis of different staple crops. The peasant association can then choose the recipe appropriate for their area, take the quantities needed, mill the ingredients themselves, mix them together and sell the mixture to their members or distribute it freely to low-income earners. Another solution is that the peasant association will decide to set aside certain fields for poor children. They can farm these as part of the communal farm. The proceeds of the harvest could then be given to the poorest families in the area. On such plots crops that are not so common, like vegetables, can be introduced.

It is likely that the demand for industrial weaning food will increase in the future, because the urban population is increasing faster than the rural population. Moreover, the 0–4 year age group will increase because the birth rate in Ethiopia is still very high. At the moment, the proportion of children under the age of four is estimated to be 17 per cent of the total population. If all these children eat 15 kg of supplementary food a year, the demand for industrial weaning food will reach about 5 million tons a year. At the moment, the annual production of Faffa Food Plants is less than 10 000 tons.

Faffa products are much in demand because their price is low and there is a shortage on the commercial market. About half the Faffa Food Plant production between 1976 and 1978 (about 8000 tons) went to the relief market, but it was planned to increase the commercial market to 6000 tons in the budget year 1979/80 and 2500 tons for the relief market. Faffa products are necessary in urban areas where most of the food is bought and for mothers who are unable or unwilling to breast-feed their children. Moreover, these products can provide supplementary feeding at times when food is not available in rural areas.

In order to reduce the subsidies of Faffa products and to create more employment and incomes within the country, the Faffa Food Plants are trying to include more local ingredients in their products. One example is a project in Eastern Wollega, where they have planted 2000 hectares of soybeans. Another example is to use maize – which is cheaper than soyflour and wheat – as an ingredient. Finally, they are experimenting with the use of sorghum and millet in weaning food.

Notes

1. University of Uppsala.
2. Eriksen and Mikkelsen, *Afrika i kartor och siffror* (Uppsala: Scandinavian Institute of African Studies, 1979).
3. Sjöström and Sjöström, *Literacy Schools in a Rural Society* (Uppsala: Scandinavian Institute of African Studies, 1977).
4. Michael Stahl, *Ethiopia: Political Contradictions in Agricultural Development* (Stockholm: Rabèn & Sjögren, 1974).
5. Negarit Gazeta, *Proclamation no. 31 of 1975, to Provide for the Public Ownership of Rural Lands* (Addis Ababa, 29 April 1975).
6. Negarit Gazeta, *Proclamation no. 47 of 1975. Government Ownership of Urban Lands and Extra Houses* (Addis Ababa, 26 July 1975).
7. Michael Stahl, *New Seeds in Old Soil* (Uppsala: Scandinavian Institute of African Studies, 1977).
8. Negarit Gazeta, *Proclamation no. 31 of 1975*, ch. 3.
9. Ibid, and Stahl, *New Seeds in Old Soil.*
10. Negarit Gazeta, *Proclamation no. 31 of 1975.*
11. Negarit Gazeta, *Proclamation no. 47 of 1975.*
12. King *et al.*, *Nutrition for Developing Countries* (Nairobi: Oxford University Press, 1972).
13. Cameron and Hofvander, *Manual on Feeding Infants and Young Children* (United Nations, 1976).
14. A report submitted to the National Revolutionary Development Campaign and Central Planning Supreme Court (August 1979), hereafter referred to as Faffa Food Plant Report.

210   *Weaning Food and Low-income Consumers in Ethiopia*

15. Wickström, *Bo in Undernäring i u-land* (Stockholm: Almguist & Wiksell, 1972); and ENI, *A Plan for Diversification of the ENI Supplementary Food Programme*, 2nd edn (Addis Ababa, Nov 1974).
16. Faffa Food Plant Report, *A Plan for Diversification* (Faffa Food Plants, 1974).
17. Ibid.
18. K. E. Knutsson, *Authority and Change* (Göteborg: Etnografiska Museet, 1967).
19. E. Harsche, *Wirtschaft und Gesellschaft in West-Wollega* (Hermannsburg: Verlag Missionshandlung Hermannsburg, 1975).
20. R. Selinus, *The Traditional Foods of the Central Ethiopian Highlands* (Uppsala: Scandinavian Institute of African Studies, 1971).
21. Ibid.
22. Cameron and Hofvander, *Manual on Feeding Infants and Young Children*, ch. 15.

9 Passenger Transport in Karachi

Mateen Thobani[1]

9.1 The Problem

Introduction

From a fishing town of less than half a million inhabitants in 1947, Karachi has grown into a thriving metropolis of 6 million. The average annual growth rate of 6 per cent since the early 1960s has resulted in increasing problems in city services, particularly passenger transport. This chapter identifies some of the major problems besetting Karachi's transport system and, based on an analysis of household data, suggests some remedies to ease the situation.

Concern in Karachi's transport system is not new. Several studies[2] have made recommendations to reduce the problems, the most notable being those by the Master Plan Office (MPO) of the Karachi Development Authority. Evolved out of a United Nations Development Programme grant to the city, the MPO has published many reports on Karachi's transport system. Most of these studies were done jointly with experts from the United Nations and two international consulting firms, PADCO and TERPLAN. Many of the recommendations made in this paper are similar to their proposals. This study strengthens some of the conclusions reached by other studies; however, it goes one step further in that it measures the effects of carrying out the proposals on the pattern of demand for various modes, and conducts a cost–benefit analysis of one of the major recommendations.

The transport sector

Buses. Buses, with their regulated fares of 25 paisa to Rs. 1.00 (100 paisa = Rs.1.00 = 10 US dollar cents) are by far the most important mode for the

211

majority of the residents. Yet this is precisely the mode of transport where the supply has lagged behind its demand over the years. Table 9.1 shows the growth of various motorised vehicles in Karachi. The data on buses, in addition to private and public city buses, include buses belonging to various organisations and businesses, inter-city buses and contract carriers. The latter are private buses typically hired by a community where commuters pay a fixed sum of money per month and are guaranteed a seat. While the total number of registered buses has been growing fairly rapidly, the number of city buses available for public transit has shown little growth.

About two-thirds of city buses are privately owned. In absolute terms, the number of private city buses decreased from 1200 in 1955 to 800 (of which 700 were on the road) in 1971. By September 1979, of the 1163 privately registered buses, about 800 were on the road. The decline in private buses since 1955 – despite rapid population growth – is primarily caused by the fare structure which has been kept very low because of political reasons. At a fare of one US dollar cent for a 3-mile bus trip in December 1974, a World Bank study found Karachi fares to be the lowest in the world. In 1956, bus fares were reduced from an average of 4 paisa per passenger-mile to $2\frac{1}{2}$ paisa per passenger-mile. In June 1979, fares averaged about 5 paisa per passenger-mile. At the time of writing they stand at about 8 paisa per passenger-mile.

In the public sector only 350 of the 600 buses owned by the public sector were on the road in September 1979. Others were unoperational because of a shortage of spare parts or owing to other repairs. While this is a marked improvement from the early 1970s, when less than 200 of a total fleet of 800 public sector buses were typically on the road, it is not surprising to find the private sector better able to maintain its buses

TABLE 9.1 Motor vehicles on the road by type (in 000)

	1971	1972	1973	1974	1975	1976	1977	1978
Cars	27	32	34	38	33	37	61	74
Motor-cycles	16	22	26	31	32	41	61	82
Taxis	3	4	4	4	3	3	5	6
Motor-rickshaws	6	6	7	7	6	7	6	6
Buses	1	1	1	2	2	3	4	5
Trucks	4	4	4	5	4	4	5	6
Others	1	2	2	3	3	5	5	8
Total	58	71	78	90	83	100	147	187

Source. Motor Vehicle Registry data.

despite low fares. Of the 350 public buses on the road in 1979, 100 were used exclusively by students at a fare of 25 paisa regardless of distance.

Bus frequencies are generally good. Most routes have a headway of less than half an hour. Based on household interviews conducted by the author in the summer of 1979, the average waiting time for a trip to work by bus was 8.5 minutes for areas within 3 miles of the centre of the city, 10.4 minutes for areas between 3 and 5 miles, and 13.7 minutes for areas further than 5 miles. However, there is considerable variation in bus intervals even on the same route. In a study conducted by the MPO[3] in 1971, the interval between buses was often twice the mean interval. This was not the result of congestion but rather the practice of waiting for the bus to fill up before leaving. Average speeds ranged from 10 to 14 m.p.h. with little variation throughout the day. (The effect of increased congestion during peak hours was compensated for by the decreased need of waiting at stops to pick up additional passengers.) In the sample for this study, the average speed for trips to work by bus was just over 12 m.p.h.

According to a large household survey conducted by the MPO in 1971, the number of trips to work by personal motorised modes such as cars, motor-cycles, taxis and rickshaws was less than 10 per cent of total trips to work. Walking and bicycling accounted for about 40 per cent, while buses, trams, rail and launch comprised about 50 per cent.

The average trip length was over 5 miles for the mass transportation and personal motorised mode categories. Walking and bicycle trips tended to be very short (less than 1 mile). In terms of passenger-miles, mass transportation accounts for 76.3 per cent of all trips to work. The main mode (86.4 per cent) of mass transportation is the bus.

Based on demand projections carried out by the MPO in 1971, the Office recommended an increase in the number of buses to 2580 by 1977. In addition, it recommended to upgrade the tram or streetcar system which has since been discontinued altogether. As noted earlier, the current bus fleet comes nowhere near to the figure projected for 1977. To some extent the frustrated demand has been picked up by the unanticipated dramatic rise in minibuses. However, there seems to be a case for increasing bus services, particularly in view of the fact that most buses are completely full and have passengers hanging out of windows and precariously balanced in doorways.

Privately owned minibuses.   Privately owned minibuses were first introduced in 1972 and meant to serve as fourteen-passenger wagons. However, by early 1979 they typically carried eighteen or nineteen passengers at regulated fares of Rs. 0.75 to Rs. 1.25 per person. Known as yellow devils or yellow deathtraps, depending on whether one dodges them or rides in them, they have helped fill a gap between overcrowded buses on the one hand and expensive rickshaws and taxis on the other.

As of September 1979, there were about 3000 minibuses registered in Karachi, of which about 12 per cent were out of service. While almost all minibuses up to early 1979 were fourteen-passenger Ford vans, the last one and a half years have seen a huge increase in the number of diesel-powered twenty-six passenger Mazda minibuses. Because of the import duty structure and the cheaper cost of diesel, both the purchase cost and operating cost of the Mazda is lower. The increased passenger capacity also makes it more desirable. With the exception of routes to areas further than 5 miles from the centre of town, the average waiting time for minibuses in the sample for this study was a little higher than for buses. Average speeds for minibuses are about 25 per cent faster than buses. They are significantly more comfortable than buses if one has a seat, but the comfort advantage is doubtful if one has to crouch standing.

Taxis and motor-rickshaws.   Although the number of taxis has doubled since the early 1970s, the number of motor-rickshaws (three-wheeled motor-cycles meant to carry two passengers in addition to the driver) remained constant up to the end of 1978. While fares for both modes are low (Rs. 0.70 and Rs. 1.00 per mile in early 1979), it is not clear why taxis have grown relative to motor-rickshaws. Part of the reason may be the growth of minibuses which has detracted more from rickshaw than taxi demand. Due to the high level of noise and smoke pollution coupled with a high risk of accidents, it is not clear whether additional rickshaws are desirable.[4]

Almost the whole taxi fleet is composed of small Japanese cars with an engine size of less than 1300 cc. Due to a significant import-duty reduction a large proportion are reconditioned cars made in the early 1970s. During peak hours it is common to find shared taxi rides organised by the taxi driver who manages to obtain twice the metered fare by charging each of four customers half the metered fare. Taxi and rickshaw drivers often charge more than the metered fare even during non-peak hours. Depending on the whim of the driver, the appearance of the customer and his intended destination, they may charge up to 20 per cent more, and may even charge ten times the amount to newly arrived foreigners.

Cars and motor-cycles.   Cars and motor-cycles have exhibited a dramatic increase over the last decade. The congestion on some major roads because of this has resulted in buses and animal-drawn vehicles being excluded from these roads. Cars, however, are not used as inefficiently (in terms of utilisation) as might be expected owing to the presence of extended families and chauffeurs. Motor-cycles also often serve as family transport. It is not an uncommon sight to see a couple with three children on a motor-scooter.

Most motor-cycles are either Vespa scooters or light motor-cycles in the 50–90 cc range. Duties (import duty plus sales tax) on motor-cycles are 123 per cent. Most cars are small Japanese cars, mainly because of the escalating duty structure. Duties range from 127.5 per cent on cars with an engine size of less than 1000 cc to 485 per cent on cars over 1600 cc. Duties on spare parts are also high, and 78-octane petrol sells for about $2.10 per US gallon (1979).

Bicycles and walking. Bicycles are not used as much as one would expect in a relatively flat city with little rain, where the locally manufactured bicycles are quite cheap. The use of bicycles is limited primarily to certain domestic or unskilled workers, as a result perhaps of the status consciousness of the white-collar workers who would prefer to ride in a crowded bus than be seen on a bicycle. Other reasons may be the strain of riding a bicycle in the hot Karachi sun and the danger of being knocked over by passing motorised vehicles. Traffic laws are flagrantly violated owing to an ineffective police force and uneducated drivers.[5]

Although a large number of trips are undertaken by foot and bicycle the average trip length is just over 1 mile for trips to work and one-fifth of a mile for social and recreational trips. To a large extent this is probably the result of the non-availability of clear footpaths. Where they exist, footpaths are often encroached upon by pedlars, hawkers, small stalls and hand carts, leaving the pedestrian to take his chance by weaving between motorised and animal-drawn vehicular traffic or choosing some other mode of travel.

Railways. There is a circular railway running from a suburb of Karachi through the main industrial site to the centre of the city. It carries over 100 000 passengers daily. While used heavily by residents of the suburb, relatively few passengers embark on its stops in the city. This is probably because of uncertain scheduling and the presence of alternative modes such as bus or minibus. According to a Master report, 'there is great scope for improvement in management and scheduling. The railway infrastructure is of good quality and the installed capacity can permit sizeable increases in traffic without further investment.'[6]

Animal-drawn vehicles.　Animal-drawn vehicles are on the decline in Karachi. From a total of 8000 such vehicles in 1971, the current total stands at less than 2000. While tongas (small horse-driven carts), donkey carts and camel carts are used primarily as a substitute for trucks, victorias (large stately horse-drawn carriages) are used exclusively for passenger transport.

9.2 Methodology

Choice of model

Since one of the primary aims of this chapter is to identify appropriate transport modes for low-income groups, any model that is selected must be able to do this and to compare the results with the existing situation. Other questions that may be of interest to the policy-maker are demand elasticities for public modes with respect to their fares and other attributes such as time taken. In Karachi, both fares and petrol prices are controlled by the government and it is instructive to model the demand for a particular mode as a function of its own fare and other fares. Thus the model would predict not only the effect of a change in bus fare on the demand for buses but on the demand for minibuses and rickshaws too. The calculation of values of travel time and waiting time would be useful both for comparison purposes and in evaluating projects that result in a decrease in travel or waiting time for a mode.

In addition to the above questions of mode choice or modal split, there are some other questions that may be of use to the policy-maker – those of mode ownership. What determines a household's decision to buy a car, motor-cycle or bicycle? How does an increase in the maintenance cost of a car influence the decision to buy a car or motor-cycle?

The questions raised in the first paragraph of this section suggest some sort of a modal split model. The more sophisticated of the earlier modal split models expresses the ratio of number of trips by car to the number of trips by public transport from one zone to another as a function of the time taken in each mode, the socioeconomic characteristics of the origin zone, and the land-use characteristics of the destination zone.[7] Although based on aggregated data, these models are behavioural and policy-oriented. However, the choice of two modes in the aggregate modal split models appears to be restrictive and would have little relevance in a less developed country which typically has a rich variety of distinct modes. In addition, variables are sometimes aggregated in ways that are not useful for policy formulation.

The newer probabilistic models remove these shortcomings. These models consider the probability of an individual making a certain choice (e.g. the choice of mode to work) as a function of attributes of the choice and socioeconomic characteristics of the individual. As such, they are behavioural, are based on disaggregated data, and are policy-oriented in that they are sensitive to changes in policy variables. In his summary paper, Stopher[8] concludes that probabilistic models are best suited to infer the value of time.

Since attributes of the model include such variables as value of time (as a function of income) and fare, the estimated model can yield demand elasticity estimates of these variables. The coefficients can then be used to see which would be the preferred mode for persons of various income groups given the characteristics of the trip and individual. A possible hypothesis to test is whether the bus is predicted to be the optimal mode for certain groups of people but is not used because of its unavailability. The probabilistic model would first be estimated under the assumption that an individual can choose a bus only if the bus is available to the person. The results of this model could then be compared with the unconstrained problem when the bus is available to all.

An important advantage of the probabilistic models is that under certain assumptions they can be derived from a theory of stochastic utility maximisation. Furthermore, the decision set may include several modes and allows for joint decisions such as the joint probability of buying a car and going to work on a bus. If the joint nature of a decision is ignored it may bias coefficient estimates. Decisions such as whether or not to make a trip or choice of time of day to make the trip can also be handled, although they are usually difficult to model or data are unavailable.

While the probabilistic models are often laid out within a general framework that allows for elaborate joint decisions, few studies actually estimate the joint decisions. Among these are Adler and Ben-Akiva's[9] study on shopping trips and Lerman's[10] study on choice of residential location, type of housing, automobile ownership and mode to work. More recently, Train[11] estimates a joint-choice model of mode ownership and mode to work based on a sample of households in the San Francisco area. Using Train's approach, this chapter estimates a joint-choice model of mode ownership and mode to work based on household data in Karachi.

There are several reasons for selecting to model this set of joint decisions. The trip to work is a large percentage of total trips made by the household (36.9 per cent of all non-home trips in our sample). It is made to a fixed destination at a relatively fixed time of day and its demand is inelastic. This allows us to ignore the effect of a change in price on the decision to go to an alternate location or a decision not to make the trip at all.

One would expect that the decision to buy a mode is determined simultaneously with the decision to take a particular mode to work. If this were indeed the case and the simultaneity ignored, coefficient estimates may be biased. In addition, by modelling the joint decision it would be possible to see the effect of, say, an increase in petrol price not only on the demand for a mode to work but also on the demand for car ownership. Using income and expenditures on maintenance and depreciation as independent variables, the mode ownership issues raised in the second paragraph of this section could also be handled.

The nested logit model

For reasons explained by the author[12] in a more extended publication, this study uses the so-called nested logit model, which is a generalised version of the multinomial model.[13]

To illustrate the model consider a simple case where a person makes the joint decision of whether or not to own a car, and whether to go to work by car, bus or by foot. If this person is not allowed to decide not to buy a car but go to work in one, he has five choices available to him as shown in the lowest level of the tree in Figure 9.1.

Under the assumptions of the multinomial logit model one could now estimate the five-alternative model directly. However, the alternative walk while owning a car and walk while not owning a car are likely to be close substitutes, thus violating one of the assumptions of the multinomial logic model. By breaking up the decision into the conditional

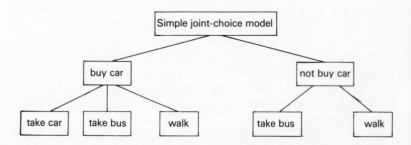

FIGURE 9.1 Disaggregation of a simple joint-choice model

probability of choosing a mode to work given the decision to buy a car, and the marginal probability of buying a car, the five-alternative model is broken up into a two-step estimation procedure. The first step is a multinomial logit model of three alternatives and the second a simple logit model. The logit structure is only imposed on the two smaller models. In the first step one only requires the three modes – car, bus and walking – to be distinct, while the second stage requires the alternative not to own a car to be distinct from owning a car.

At this stage it is worth commenting on why it was necessary to select a relatively complex estimation technique. Using ordinary least squares (OLS) is not suitable for such a probabilistic model for several reasons. Since the dependent variable is a probability it is bounded by 0 and 1 whereas OLS predictions could well lie beyond these bounds.

These problems are avoided by both probit and logit. The decision to use logit rather than probit is based mainly on computational grounds. Logit is significantly cheaper in the case of polytomous variables.

The log likelihood function is one of the best indicators as regards an improvement in the model. Two goodness-of-fit measures are based on it. One such measure is the likelihood ratio statistic:

$$-2\left[L(\beta_0) - L(\tilde{\beta})\right]$$

which, under the null hypothesis that all parameters equal zero, is asymptotically distributed as a chi-squared variate with k degrees of freedom, where k is the number of estimated parameters. $L(\beta)$ is the log likelihood evaluated alternatively at $\tilde{\beta}$, the maximum likelihood estimate of the parameter and at β_0, a vector of zeros. If the statistic is large it leads us to reject the null hypothesis.

Another goodness-of-fit statistic is the likelihood ratio index defined as:

$$\rho^2 = 1 - \left[L(\beta)/L(\beta_0)\right]$$

The index is analogous to the OLS multiple correlation coefficient.

9.3 Description of the Karachi Model

The survey

Four hundred households throughout Karachi were interviewed for the purpose of this study. A description of the sampling methodology is given below.

In order to ensure enough variation in mode choice, it was important to have sufficient observations within various income categories. Accordingly, each homogeneous income area in Karachi was assigned to one of five income categories based on their median household income.[14]

Similarly, it was considered important to have sufficient observations on households that live near the centre of the city as well as those that live at a distance. Therefore these homogeneous income areas were further split into three distance categories – 0–3 miles, 3–5 miles, and 5 or more miles from the centre of the city. Certain areas were then discarded for special reasons such as for being a red-light district or for having moved significantly from its income group since the UNDP study. To decrease the probability that one of the chosen areas was not a representative member of the cell, two areas were randomly selected from the list of areas in each cell. Since random sampling within these relatively large areas was not feasible, a sample of about fifty households within each of the areas was chosen by the author after an inspection to ensure that the block of houses thus selected was not atypical for the area. Interviewers were then sent out to interview at least thirteen households in each of the thirty areas selected.

The richness of the data comes from information not only on the trips made the day earlier by each individual of the household, but also on demographic characteristics of the household, income by source, hours worked, expenditure by various categories, assets, availability of public modes, present and purchase values for owned modes and maintenance expenditures on these modes. The questions on trips made included origin and destination, purpose, mode used, in-transit and waiting time, distance covered, number of persons in mode, cost (if public mode), and trip frequency.

The survey shows that over one-fifth of the trips are work trips. Not surprisingly for Karachi, social trips to visit friends or relatives account for over 10 per cent of total trips or 18.6 per cent of all non-home trips. The home category is not very revealing as it does not distinguish between coming home from work and coming home from some other destination. The other category includes trips made for religious, tutorial or training purposes. Due to the manner in which the sample was selected and because households with cars tend to make more trips, there were about equal numbers of trips by the two principal modes – car and bus. Although a good proportion of the sample included persons in the below Rs.1600/month income category who were most likely to own bicycles, relatively few bicycle trips and still fewer bicycle owners were found in the sample. This seems to support the hypothesis advanced earlier that bicycles are underutilised. Unfortunately, the paucity of bicycle owners also makes statistical analysis of bicycle ownership extremely unreliable.

Table 9.2 shows that the car is the predominant mode of transport for high-income groups while the bus, minibus and walking are the principal transport for low-income classes.

Table 9.3 indicates that the buses and minibuses are, on average, available to about three-quarters of all households. This proportion is somewhat lower for the low- and high-income groups, but this situation is more serious for low-income groups since they have no alternative modes of transportation. In other words, there is likely to be a frustrated demand for buses in the low-income areas. The relatively high availability of minibuses in the low-income areas is probably the result of the unavailability of buses in those areas.

Another feature of the Karachi transport system is the high number of persons transported in each trip. It is interesting to note for example that the frequency of trips with two persons on a motor-cycle is almost as high as that of one person, while there were more trips with two persons

TABLE 9.2 Percentage of trips made in each mode by income group

Rs. ('000)/month	0–0.8	0.8–1.6	1.6–3.0	3.0–5.0	5.0+
Car	0.5	3.9	15.3	37.9	65.7
Motor-bike	1.0	5.2	12.2	12.8	7.2
Bicycle	3.5	3.6	2.3	1.6	1.5
Taxi	0.1	2.5	1.8	2.4	3.3
Rickshaw	2.0	7.7	3.7	3.5	2.1
Minibus	13.6	10.5	11.6	4.7	2.8
Walking	30.6	19.6	16.1	11.0	5.6
Bus	48.7	47.0	37.0	26.1	11.8
Total	100.0	100.0	100.0	100.0	100.0

TABLE 9.3 Bus and minibus availability by income group (in % of households)

Rs. ('000)/month	0–0.8	0.8–1.6	1.6–3.0	3.0–5.0	5.0+	Total
Availability of:						
Bus	66.7	74.3	89.0	73.9	67.9	75.2
Minibus	70.4	72.9	79.5	69.2	66.0	72.1
(Households)	17.1	22.2	23.2	20.6	16.8	100.0

to a car than with one person to a car. The distribution of number of persons in a minibus (not shown) for work trips reflects the scarcity of adequate public transportation in Karachi. Of the total of thirty-nine work trips by minibus for which data on number of persons in vehicle existed, twenty-six trips had more than fourteen persons (fourteen is the seating capacity). An additional nine persons responded that it was completely full which is taken to mean that there was no seat available. Nine persons responded that there were twenty persons in the vehicle.

In order to get a feel for the non-monetary factors that influence travel demand behaviour, an attitudinal survey was carried out that attempted to quantify individual preferences for comfort, danger, physical strain, privacy and status. It is interesting to note some factors that emerged during the course of pre-testing. It was soon apparent that interviewees could not understand an abstract concept such as 'willingness to pay' for additional comfort, etc. Therefore the questions were rephrased in terms of more concrete situations of the sort 'how much would you be willing to pay a seated person to obtain his seat?' or 'how much would someone have to pay you give up your seat?'. There was violent opposition to these sorts of questions on grounds of 'what kind of a person do you think I am anyway?' When the questions were reworded in the form of 'how much extra would you be willing to pay to be guaranteed a seat on the bus?', they felt that the survey was a ploy to raise bus fares without a commensurate increase in service. Finally it was decided to pose questions based on hypothetical situations where an individual had to make a decision of trading minutes of leisure time with obtaining a seat on a minibus. This approach worked rather well in that answers were readily forthcoming and a wide range of variation was observed in the distribution of responses.

The alternatives and variables

Although it would have been desirable to consider the choices of mode ownership to be all possible combinations of buying a car, motor-cycle or bicycle, this was not possible because of a lack of data. There were very few people in the sample who made both a work trip and owned either a bicycle or more than one mode. Therefore it was decided to consider only three alternatives of mode ownership – own a car, own a motor-cycle or own nothing. Other observations were discarded from the sample.

The modes to work for a consumer include car, motor-cycle, bicycle, taxi, motor rickshaw, minibus, walking and bus. The total number of choices being considered is therefore 24. However, in order to exclude the possibility of a person not owning a car or motor-cycle but going to work in one, the choice set can be reduced to 20. It will be assumed that a car is available to an individual for his work trip if the household owns a car but that a motor-cycle or bicycle is available to the individual only if the individual is himself the owner. This distinction is necessary because many cars are owned by the firm for tax purposes. Also it is likely that the person going on the work trip has priority on the car. On the other hand it is not realistic to expect that the earner will take his college-going son's motor-cycle to work. The choice set may be further reduced for some individuals as certain public modes may not be available in certain areas. Due to the lack of observations on bicycle owners, a bicycle is treated like a public mode. Either it is available or not – the consumer does not have the option to buy one. The bicycle is available to only those people who own one or ride to work on one.

There are three types of variables in the model – those that vary with mode choice to work such as in-transit time, those that vary with mode ownership such as depreciation cost, and those that vary with individuals such as income. Data are only observed for chosen alternatives. Data on attributes of alternate choices have to be generated from these observed data.

The dependent variable for the modal split analysis or first step of the estimation process, is the probability of the consumer choosing a certain mode given the availability of the mode. For the choice of ownership of mode, the dependent variable is the probability of owning a car, motor-cycle or no modes.

A description of the way by which the independent variable (x_i) for the modal split analysis (bottom level of Figure 9.1) were generated is given below:

1. Value of waiting time (WWAIT): The traditional way to treat the value of waiting time is that it equals the waiting time times the wage rate. The wage rate is defined in rupees per minute. The waiting time for a car, motor-cycle, bicycle or walking was assumed to be zero while that for modes not selected was taken to be the average waiting time in minutes for each mode. This is likely to be an underestimate because of a sample selectivity problem – the choice not to use a bus may be motivated by the excessive waiting time for buses in the consumer's area. For buses and minibuses, since there were enough observations, it was possible to take the average time in the three distance categories and so reduce the sample selectivity problem.

2. Value of in-transit time (WTRAN): This variable is the product of in-transit time and the wage rate. The actual in-transit time in minutes was used when the alternative was selected. In order to generate the in-transit time for the alternatives not selected, a simple regression of distance (in tenths of a mile) on in-transit time was carried out for work trips by each mode. It was necessary to suppress the constant term as trips of zero distance are expected to take zero minutes. Including the constant term causes short trips to have large percentage errors.

3. Marginal monetary cost (COST): This refers to the fare if the mode is a public mode. For cars and motor-cycles it is the petrol cost of the trip divided by the number of persons (excluding the driver) in the car. This is not a bad assumption since parking is free virtually everywhere and the addition to depreciation and maintenance cost caused by the work trip is likely to be negligible as they tend to be related to time rather than mileage. To generate data on per mile fares of public modes when the alternative was not selected, it was decided to resort to regression analysis rather than using official fares. Not surprisingly, the regression of distance on rickshaw yielded a coefficient that was 20 per cent higher than the official fare of Rs.0.70 per mile. Observations on taxi fares to work were too few to give good results and so a premium of 20 per cent over the official fare was assumed. Bus and minibus coefficients were similarly generated with all coefficients being significant at the 1 per cent level. Data on the cost of car trip to work when the car was owned was based on the interviewee's estimate of the mileage per gallon. When two or more of a mode were owned the information on the first was used.

4. Dummy for long walking trips (DWALK): Since walking is likely to be more onerous than other modes, a dummy for walking is constructed which is a function of distance. It equals zero unless the trip is longer than 1 mile in which case it is one. Alternate specifications of DWALK (proportional to and exponential with distance) were also tried but were not as satisfactory.

5. Dummy for car use (DHEAD): While everyone in the household is assumed to have a car available for the trip to work if anyone in the household has a car, the head of household is more likely to get the car. DHEAD, which is one when the car is available and the individual is the head of household, and zero otherwise, captures this effect.

6. Opportunity costs of time for near and distant households (NEARWW, FARWW, NEARWR, FARWR): If individuals have different values of time, then other things being equal, the ones with lower values would tend to locate further from the centre of city. These variables allow one to test for this. FARWW and FARWR equal

WWAIT and WTRAN respectively if the individual is located more than 5 miles from the centre of the city, and zero otherwise. Conversely, NEARWR and FARWW equal WWAIT and WTRAN when the individual lives within 5 miles of the centre, and zero otherwise.

The variables (y_c) used in the mode ownership analysis are described below:

1. Aggregate utility from the trip to work or inclusive value (INCLUS): For each individual, this variable reflects the aggregate utility from the trip to work in the three mode-ownership categories. The definition of inclusive value can be found in a more extended publication on this subject by the author.[15]

2. Mode expenditures as function of income (EXPINC): The sum of maintenance expenditures and depreciation costs on a car or motor-cycle is taken to be the cost of owning a mode. The actual values of yearly maintenance cost in thousands of rupees obtained from detailed questions on mode servicing, tuning, spare parts, labour, body work, tyres, batteries, insurance and overhauling were used when the alternative was owned. Depreciation costs were calculated by taking the interviewee's estimate of the purchase value and present value of the vehicle discounted for inflation and dividing by the number of years the vehicle had been in his possession. The actual values of average yearly depreciation in real thousands of rupees were used when available. Imputed values for the expenditure variable were simply taken to be the average values of maintenance and depreciation costs. In order to account for the decreasing marginal utility of money, the expenditure variable was divided by total household income from both earned and unearned sources in thousands of rupees per year.

3. Dummies for living close to the centre of city (NRCAR, NRBIKE): These variables test the hypothesis that individuals living close to the centre of the city obtain a lower utility from owning a car or motor-cycle. NRCAR equals one if the alternative is car and the individual lives within 3 miles of the centre, and zero otherwise. Similarly, NRBIKE equals one if the alternative is car and the individual lives within 3 miles of the centre, and zero otherwise.

4. Dummies for being a nuclear family (NUKCAR and NUKBIK): The construction of these variables allows one to test the hypothesis that nuclear familiar families are less likely to own a car. NUKCAR equals one if the family is composed of only the head of household, spouse or children and the alternative is one, and zero otherwise. NUKBIK was constructed in a similar manner for the second alternative.

5. Number of people in the household (NMEMCR and NMEMBK): NMEMCR equals the number of members in the household over five years of age if the alternative is car, and zero otherwise. Similarly NMEMBK equals the number of members in the household in the second alternative. The signs of these variables are expected to be negative because of the hypothesis that larger households would have less money left over to buy a car or motor-cycle (assuming that the nuclear family dummies adequately capture the effect of extended families being able to use the mode more efficiently).

9.4 Results

Modal split analysis

The estimation of the model was carried out using a statistical package QUAIL, developed by Wills, Glanville and McFadden at the University of California.

The coefficient estimates and other relevant statistics for four alternate specifications of the first step of the joint-choice model are given in Table 9.4. The restriction that the value of time is proportional to the wage rate is implicitly imposed on each of the models. Model A considers only the time and cost attributes of each mode. In order to account for the additional disutility of walking long distances, Model B introduces a dummy DWALK. In addition, Model B also uses a dummy DHEAD to test for the higher probability of the head of household having access to the family car. Model C employs two alternative specific dummies DMINI and DBUS to capture some of the negative attributes of non-personal modes. Finally, in order to test the hypothesis that people who locate further from the centre of city have lower values of time, the variables FARWW, NEARWW, FARWR and NEARWR replace WWAIT and WTRAN of Model B.

With the exception of NEARWR and FARWR, all coefficients are significant at the 1 per cent level. The likelihood ratio statistic which is distributed as chi-square leads one to reject the hypothesis that all coefficients are zero. Models B, C and D all predict equally well, with Model C yielding the best fit as reflected in its higher likelihood ratio index of the observed trips by various modes; Model C correctly predicts 81.5 per cent.[16]

Since these coefficients are parameter estimates of a linear utility function, their ratio is the marginal rate of substitution. The ratio of the

TABLE 9.4 Coefficient estimates–modal split

Variable		Model A	Model B	Model C	Model D
WWAIT	coeff.	− 1.163	− 1.044	− 0.8026	
	t-stat	− 9.835	− 7.341	− 5.590	
WTRAN	coeff.	− 0.4994	− 0.1293	− 0.1034	
	t-stat	− 9.087	− 6.184	− 4.489	
COST	coeff.	− 0.3784	− 0.5763	− 0.6824	− 0.5200
	t-stat	− 5.783	− 7.167	− 7.398	− 6.511
DWALK	coeff.		− 3.729	− 4.358	− 4.322
	t-stat		− 8.234	− 8.676	− 7.415
DHEAD	coeff.		1.643	1.629	1.591
	t-stat		3.016	2.946	2.901
DMINI	coeff.			− 1.356	
	t-stat			− 4.918	
DBUS	coeff.			− 0.5540	
	t-stat			− 2.325	
FARWW	coeff.				− 0.7055
	t-stat				− 3.926
NEARWW	coeff.				− 1.395
	t-stat				− 6.662
FARWR	coeff.				− 0.0955
	t-stat				− 1.690
NEARWR	coeff.				− 0.0445
	t-stat				− 0.7977
$L(\beta_0)$		− 507.6	− 507.6	− 507.6	− 507.6
$L(\beta)$		− 314.8	− 253.1	− 238.6	− 247.7
likelihood ratio stat		385.7	509.1	538.1	519.9
likelihood ratio index		0.3799	0.5014	0.5300	0.5121
% predicted correct		70.22	79.49	81.46	79.49
degrees of freedom		1 194	1 192	1 190	1 190

coefficients of WWAIT to WTRAN in Model A of 2.3 implies that the marginal disutility of waiting time is 2.3 times that of in-transit time. Most studies that distinguish between time spent waiting for a mode and time spent travelling in the mode find this ratio to be in the range 2 to 3. In Model B, the variables WTRAN and DWALK are highly correlated since walking is the slowest mode. I suspect, therefore, that the resulting collinearity attributes some of the disutility of WTRAN to DWALK leading to a low absolute value for WTRAN. The same problem exists in Models C and D. Although Models B, C, and D do not provide us with a good estimate of the coefficients of DWALK or the in-transit time variables, the other coefficients are reasonable and the models have a high prediction which make them useful in subsequent simulations.

Note also that Models B and C are more general versions of Model A, and the null hypothesis that the coefficients of DWALK and DHEAD are zero is rejected at the 1 per cent level.

As explained in section 9.2, the marginal rate of substitution between an attribute (e.g. WWAIT) and money is the ratio of the coefficient of the attribute to the coefficient of price. Therefore the value of WWAIT is 3.07 (1.163/0.3784) in Model A, and hence the value of waiting time is 3.07 times the wage rate. (Recall that WWAIT equals the wage times the waiting time.) Similarly the value of in-transit time is 1.32 times the wage rate. This is higher than expected and could be the result of the absence of some variables such as comfort and convenience which are negatively correlated with waiting time and in-transit time, and strain and danger which are positively correlated with the time variables.[17]

The dummy for walking deals with the above problem, decreasing the disutility of waiting time to 2.38 times the wage rate. The smaller figure of 0.22 (0.1293/0.5763) for WTRAN is the result of the collinearity problem mentioned earlier. Adding the alternative specific dummies for buses and minibuses (Model C), the value of waiting time reduces to 1.2 times the wage rate.

In Model D, the value of waiting time for people who live close to the centre of the city is revealed to be 2.68 times the wage, while for those who live far away it is 1.36 times the wage. This result of the value of time being different depending on the proximity to the centre of town was also borne out by Train.[18] While coefficients for in-transit time are not significant for near and distant households, an appropriately constructed chi-square test shows that the hypothesis of the coefficients of both NEARWR and FARWR being simultaneously zero is rejected at the 1 per cent level.

In order to obtain demand elasticities for the modes with respect to the attributes it was decided to simulate the results rather than resort to a mathematically derived equation. This is because elasticities obtained from such an equation are very sensitive to where they are evaluated. To evaluate the elasticity at the mean makes little sense when there is a wide variation in values. One could evaluate at the observed values for each individual and take the mean of the sum of individual elasticities. However, the simulation procedure gives greater accuracy.

While Model C had the best fit, it was not used for simulating elasticities since it was not clear what factors were incorporated in the alternate specific dummies, and whether the dummies would be sensitive to the proposed changes. Therefore it was decided to employ Model B, which, with fewer variables, predicted almost as well as Model C.

The elasticities (Table 9.5) are based on a 10 per cent change in the independent variables. To obtain the demand elasticity of waiting time for a bus (WAIT (8)), for example, the waiting time for each trip in the bus alternative was decreased by 10 per cent. The new pattern of predicted demand (probability sum) was compared to the predicted demand before the change. The difference in the probability sums (divided by ten) yields the relevant elasticities. For convenience, only the absolute values of the elasticities are given. All elasticities have the expected sign.

Elasticities with respect to waiting time turn out to be much higher than those with respect to cost. All cross-elasticities with respect to cost are less than 0.12 while all waiting time cross-elasticities are more than 0.12. Not surprisingly, the cross-elasticities between minibuses and buses were higher than between other modes.

The last two rows show the effect of changing the waiting time and cost respectively of both buses and minibuses simultaneously. Decreasing waiting time for buses and minibuses has a significant effect on all but the car alternative. However, changing the cost of buses and minibuses simultaneously has a significant impact on car demand.

The implication of the higher waiting time elasticities suggests that more gains are likely to come as a result of decreasing waiting time for public modes (adding more minibuses or buses) than from making public transport cheaper. While changing waiting time for a bus or minibus has little impact on car demand, changing the cost of a bus or minibus has some impact. The costs and benefits to carrying out these recommendations are given in the next section.

The cross-tabulations of Table 9.3 suggest that there are several households who do not use the bus or minibus because of its unavailability. In order to test for this, the predicted probabilities of Model B, which were estimated under the assumption that a person could take the bus only when it was available, were compared with those when the mode was available to all. The results (Table 9.6) suggest that there is a 17.8 per cent frustrated demand for buses or a 30.1 per cent frustrated demand for minibuses predicted by Model B. If both minibuses and buses are available to all residents, the model predicts a 17 per cent increase in minibus use and a 6.6 per cent increase in bus use. To test the robustness of the result, Model C was also simulated under the same conditions. The results were similar to Model B, thus building a case for increasing bus and minibus service.

TABLE 9.5 Values of simulated elasticities for selected variables – first step (independent variable changes by 10%)

Mode	Car (1)	Motor-cycle (2)	Bicycle (3)	Taxi (4)	Rickshaw (5)	Minibus (6)	Walk (7)	Bus (8)
WAIT (8)	0.057	0.161	0.125	0.231	0.208	0.297	0.144	−0.401
WAIT (6)	0.041	0.079	0.100	0.369	0.171	−0.673	0.131	0.215
COST (8)	0.010	0.036	0.042	0.061	0.059	0.103	0.041	−0.117
COST (6)	0.014	0.032	0.042	0.117	0.089	−0.362	0.065	0.134
COST (1)	−0.100	0.006	0.037	0.047	0.051	0.046	0.035	0.039
COST (2)	0.001	−0.110	0.000	0.006	0.006	0.013	0.004	0.020
WAIT (6, 8)	0.096	0.236	0.222	0.583	0.369	−0.359	0.281	−0.187
COST (6, 8)	0.243	0.069	0.110	0.179	0.167	−0.261	0.107	−0.017

TABLE 9.6 Percentage change in demand if public transit were available to all

| Mode | Model B | | | Model D | | |
	Bus	Minibus	Both	Bus	Minibus	Both
Car	−3.45	−2.40	−4.77	−4.49	−3.24	−6.32
Motor-cycle	−3.93	−4.56	−7.43	−4.14	−4.88	−7.76
Bicycle	−1.70	−0.06	−1.76	−1.48	−0.01	−1.49
Taxi	−9.49	−5.70	−12.84	−9.65	−6.63	−14.49
Rickshaw	−16.41	−11.16	−21.58	−16.03	−11.56	−21.78
Minibus	−10.27	30.10	16.96	−9.45	32.44	19.45
Walk	−8.98	−5.81	−12.76	−9.15	−6.40	−13.50
Bus	17.83	−8.97	6.57	18.98	−9.71	7.46

Mode-ownership analysis

The second step of the analysis is a multinomial logit analysis of the decision to buy a car, motor-cycle, or no mode. One of the independent variables is inclusive value (INCLUS) which is the aggregate utility from the trip to work given that the individual owns a car, motor-cycle or no mode respectively. It is this variable that provides the link with the decision to choose a certain mode to work. As pointed out earlier (section 9.2), the joint-choice model is valid only if the coefficient of inclusive value lies between zero and one.

In order to test for the sensitivity of the coefficients, the model was estimated for two specifications of the first step, Models B and D (Table 9.7). All goodness-of-fit statistics support the model structure. In each case the coefficient of inclusive value lay between zero and one, indicating that the decision to choose a mode to work and the decision to buy a mode are indeed made simultaneously.

Since the coefficient of the inclusive value (INCLUS) is different from one, it suggests that the structure of the joint-choice model is not multinomial logit. On the other hand, since its value is not zero either, estimation of the mode ownership without the inclusive value variable would have yielded biased coefficients. In order to see the degree of bias, the model was re-estimated without the inclusive value term (model Z). The results (see Table 9.7) show that, while all coefficients are biased upwards, the magnitude of the bias is small.

The mode-ownership analysis also reveals that along with the costs of owning a mode, household location and demographic features play an important role. The negative coefficients for NRCAR and NRBIKE indicate that a person who lives within 3 miles of the city centre derives

TABLE 9.7 Coefficient estimates – mode ownership

Variable		Model B	Model D	Model Z
INCLUS	coeff.	0.5194	0.6521	
	t-stat	4.985	5.097	
AREACR	coeff.	− 1.407	− 1.177	− 0.9175
	t-stat	− 3.232	− 2.777	− 2.546
AREABK	coeff.	− 0.7635	− 0.5346	− 0.4184
	t-stat	− 1.601	− 1.141	− 1.104
NUKECR	coeff.	− 0.3237	− 0.3259	− 0.2669
	t-stat	− 0.8085	− 0.8106	− 0.7808
NUKEBK	coeff.	− 1.662	− 1.725	− 1.617
	t-stat	− 3.561	− 3.683	− 3.551
EXPINC	coeff.	− 3.027	− 3.035	− 2.798
	t-stat	− 4.473	− 4.423	− 5.052
NMEMCR	coeff.	− 0.0523	− 0.0714	0.0141
	t-stat	− 1.311	− 1.732	0.4309
NMEMBK	coeff.	− 0.2738	− 0.2916	− 0.2036
	t-stat	− 6.227	− 6.484	− 5.305
$L(\beta_0)$		− 338.4	− 338.4	− 369.1
$L(\tilde{\beta})$		− 163.9	− 162.7	− 202.5
likelihood ratio stat		349.0	351.3	333.3
likelihood ratio index		0.5146	0.5192	0.4515
% predicted correct		84.4	84.1	77.4
degrees of freedom		608	608	665

less utility from owning a car or motor-cycle than a person who lives further than 3 miles. The hypothesis that nuclear families are less likely to buy a car or motor-cycle is supported by the negative coefficients of NUKECR and NUKEBK. One of the most robust, the coefficient of EXPINC, indicates that the disutility of expenditures on a car or motor-cycle is a significant factor in the decision to buy a mode. The negative coefficient of NMEMCR and NMEMBK suggests that larger families have less left over from their income to afford a car or motor-cycle.

Elasticities of mode ownership with respect to household size, household income, and expenditures on cars and motor-cycles were simulated using the coefficients of Model B. The results (Table 9.8) suggest that income elasticity of demand for car ownership is high while that for motor-cycle ownership is low. Similarly, an increase in depreciation and maintenance expenditures affects car ownership significantly but barely affects motor-cycle ownership. The size of the family does not affect car ownership to any great degree but has great impact on motor-cycle ownership.

TABLE 9.8 Simulated elasticities for selected variables –
second step

	Own car	Own motor-cycle	No ownership
HHSIZE	−0.05	−1.11	0.14
INCOME	0.68	−0.02	−0.12
EXPCAR	−0.62	0.16	0.10
EXPMBIK	0.01	−0.15	0.02

9.5 Policy Analysis and Conclusions

In interviews, both consumers and transport authorities voiced the need for expanding bus services. There is much evidence that suggests a shortage of bus services – overcrowded buses, a relatively high elasticity of demand for buses with respect to waiting time, a high value of waiting time, the fact that many people in poor areas do not have a bus available to them, and simulations of the model which reveal a frustrated demand for buses at current fares and frequencies. Some obvious recommendations to improve the bus service include reducing the turnaround time of buses in the workshop and allocating routes more efficiently by means of a new origin and destination survey. However, given the present state of efficiency, would a scheme to increase the number of buses be cost-effective?

The major benefit to consumers from an increase in the number of buses is the savings from decreased waiting time for a bus. Here it is assumed that buses are allocated to routes in proportion to their frequency. Hence, a 10 per cent increase in buses results in a 10 per cent decrease in waiting time along each route. Since the long waiting times for trips to work often occur because buses are too crowded to board, the 10 per cent increase in the bus service may reduce waiting time for a bus during peak hours by more than 10 per cent. Note also that the 10 per cent figure implicitly assumes that the same proportion of the additional buses will be off the road for repairs.

The actual gains will be greater if buses are allocated to routes where they are most needed. Additional benefits include an increase in comfort and, in the long run, decreased congestion as people switch from other modes to buses.[19] Because of the difficulty in measuring the latter two forms of benefit, it was decided to use only the gain from decreased waiting time.

The benefit to a person from the decreased waiting time clearly is zero if he never takes the bus, and equals the value of his waiting time (which is equal to the wage rate times the waiting time) if he always takes the bus. What if the model predicts that a person takes the bus with probability one-half? The logical way to evaluate his gain from the decreased waiting time is to use one-half the value of his waiting time. This approach is followed here.

Using this method and parameter estimates from Model B, the rupee benefits (weighted by initial probabilities) were calculated for each work trip, and the sample sum for the five income group categories was obtained (Table 9.9). A figure for total trips to work in 1971 was obtained from the Master Plan Survey.[20] Under the assumption that trips to work are proportional to population and based on income-distribution data and population growth to 1979, estimates for the total number of trips to work in 1979 were obtained for each income group. The rupee benefit per trip for each income group was multiplied by the total trips made per day by each income group, to obtain the total rupee benefit to consumers. This figure was multiplied by twenty-five (Karachi has a six-day working week) to obtain the total monthly benefit to consumers of Rs.3.2 million.

This figure only reflects the gains from trips to work. There will be additional gains from non-work trips. Since the additional buses would be used primarily during peak hours, and since estimates for the value of time for non-work trips were not available, only gains for trips to work have been calculated.

Detailed operating cost figures for a bus were not available from the public sector. It is estimated, however, that the operating costs of 176

TABLE 9.9 Daily gain to consumers caused by decreasing waiting time for a bus by 10 %

| | Income group | | | | | |
	I	II	III	IV	V	Total
Sample gain (Rs.)	1.132	3.316	4.984	3.363	2.633	15.43
Trips in sample	54	70	73	65	53	315
Gain/trip (Rs.)	0.0210	0.0474	0.0683	0.0517	0.0497	0.0489
Total trips ('000/day)	604	1 618	293	253	132	2 900
Total gain (Rs./day)	12 950	76 650	20 000	13 090	6 560	129 250

extra buses would represent a total monthly expenditure of about Rs.2.4 million.

With a monthly benefit of Rs.3.2 million to consumers, this analysis indicates that the project is worth while. It is equivalent to saying that an individual lump-sum tax subsidy scheme could be devised to finance the expenditure on buses which would lead to a Pareto-superior position.

With more reliable and detailed cost data, a more sophisticated cost–benefit analysis could have been attempted – one that would weight the benefits to various income groups and consider the marginal social costs of each component of cost in terms of a common yardstick (e.g. rupees of public income). Given the incomplete data and the problems involved in weighting different income groups and obtaining the various parameters, it was decided to stay with comparing rupees of consumer income with private operating costs of a bus.

As an alternative to the public sector supplying the additional buses, it was considered worth while to seek a self-financing scheme that would provide a net benefit to consumers. For instance, would a 20 per cent bus fare increase coupled with a 10 per cent decrease in waiting time leave the consumers better off? Would the 20 per cent bus fare increase call forth at least 10 per cent more buses from the private sector? In short, would a 20 per cent fare increase be privately profitable to result in at least 10 per cent more buses, as well as be socially preferable to consumers who gain 10 per cent decreased waiting time at the expense of the higher fare?

The answer to the second question is uncertain. However, based on past responsiveness of the private sector to changes in fare, and the zeal with which private bus owners are clamouring for a fare increase, it is likely that the required 176 buses would be available. To see the plight of

TABLE 9.10 Daily gain to consumers caused by decreasing waiting time for a bus by 10 % and increasing fare by 20 %

| | Income group | | | | | |
	I	*II*	*III*	*IV*	*V*	*Total*
Sample gain (Rs.)	−0.035	0.478	1.34	1.54	1.76	5.08
Trips in sample	54	70	73	65	53	315
Gain/trip (Rs.)	−0.0006	0.0068	0.0184	0.0237	0.0332	0.0161
Total trips ('000/day)	604	1 618	293	253	132	2 900
Total gain (Rs./day)	−400	6 287	3 070	3 414	2 506	14 877

the consumers after the suggested scheme was implemented, the change in utility for each consumer is weighted by the initial probability and multiplied by the marginal utility of income. Sample sums for the five income group categories reveal that only the poorest group suffers a loss in utility (Table 9.10). The loss is very small as compared to the gain by the other groups. A loss in utility is experienced in 80 of the 315 trips. The loss in almost all cases is very small.

Notes

1. Mateen Thobani is currently an economist at the World Bank. This chapter is a condensed version of his Ph.D. thesis at Yale University. He would like to thank his advisers, John Quigley and T. N. Srinivasan, for their help and guidance.
2. See Esesjay Consult Limited, *Analysis of Existing Profile of Sind Road Transportation Corporation and a Short-term Augmentation of Facilities and Fleet Project* (Karachi, 1975).
3. See Master Plan for Karachi Metropolitan Region, *Final Report on Transportation* (Karachi, 1975; doc. MP-RR/94).
4. It is curious to note that with an increase in fares of public modes in July 1979, the number of registered rickshaws jumped from 6000 to 14 000.
5. The reader is referred to the Karachi Master Plan's *Final Report on Transportation* for a more elaborate description of the demand and supply characteristics of Karachi transport in 1971–3.
6. Master plan for Karachi Metropolitan Region, 'Transportation: Its Challenges to Karachi Metropolis', *Newsletter Monthly* (Karachi) (Nov 1979) p. 46.
7. See M. Fertal, E. Weiner, A. Balik and A. Sein, *Modal Split* (Washington, D.C.: United States Department of Commerce, Bureau of Public Roads, 1966) for description of the conventional approaches to modelling urban modal split.
8. P. Stopher, *Value of Travel Time*, Transportation Research Record, no. 587 (Washington, D.C.: National Academy of Sciences, Transportation Research Board, 1977).
9. T. J. Adler and M. Ben Akiva, *Joint-choice Model for Frequency, Destination, and Travel Mode for Shopping Trips*, Transportation Research Record, no. 569 (Washington, D.C.: Transportation Research Board, 1976).
10. Steven R. Lerman, *Location, Housing, Automobile Ownership, and Mode to Work: A Joint Choice Analysis*, Transportation Research Record, no. 610 (Washington, D.C.: Transportation Research Board, 1977).
11. K. Train, 'A Structural Logit Model of Auto Ownership and Mode Choice', *Review of Economic Studies*, vol. 48 (1980).
12. T. Domencich and D. McFadden, *Urban Travel Demand: A Behavioural Analysis* (Amsterdam: North-Holland, 1975); D. McFadden, *Modelling the*

Choice of Residential Location, Cowles Discussion Paper, no. 477 (New Haven, Connecticut: Yale University, 1977).

13. M. Thobani, 'Passenger Transport in Karachi: A Nested Logit Model', Ph.D. thesis (New Haven, Connecticut: Yale University, 1981).

14. Such a study was carried out by a UNDP mission to the City (1972). In addition to income, the groups were homogeneous with respect to level of literacy, water and sewerage connections, and percentage of brick or concrete ('pucca') houses.

15. See Thobani, 'Passenger Transport in Karachi', p. 30.

16. If all trips were equally likely, a random method would be expected to predict correctly 25.4 per cent.

17. Note that in this joint-choice model the use of a dummy for car to capture the effects of comfort and convenience results in being a dummy for car ownership since non-car owners do not have the option to take a car.

18. Train, 'A Structural Logit Model 7, Auto Ownership and Mode Choice'.

19. For the same road space, buses carry passengers more efficiently. Since the introduction of buses would typically affect travel times in all modes, it is not clear what the end result will be. However, it is likely that the first-order effect of buses being more efficient in carrying passengers will outweigh any second-order effects.

20. See Master Plan for Karachi Metropolitan Region, *Final Report on Transportation.*

10 Conclusions and Policy Implications

Wouter van Ginneken and Christopher Baron

This chapter endeavours to draw whatever general conclusions are possible from the case-studies of the various countries in the eight preceding chapters. In so doing we will endeavour to gather together evidence from the studies responding to the hypotheses set out in Chapter 1. In the second part of the chapter some suggestions are presented on potentially fruitful areas for further research. The third and final part examines the implications of the case-studies for the policies which might be adopted by governments, trade unions, employers' organisations and consumer associations in developing countries. In these countries appropriate product choice may have an especially significant contribution to make to the alleviation of poverty and the generation of employment opportunities.

These tasks are by no means easy, despite the precision of many of the individual studies in this book. As a group, the studies are disparate, covering a range of products and employing different methodologies of analysis. Of the countries in which studies were carried out four were in Asia, including Pakistan, Malaysia, Bangladesh and India; three were in African countries, namely Kenya, Ethiopia and Ghana. No case-study was carried out in Latin America, although Barbados was the subject of the chapter by James on soap. We may conclude that the studies are reasonably balanced, at least between the two poorest developing continents, Africa and Asia.

The products included are diverse. One concerns weaning (baby) foods. Personal non-food articles included are soaps (in Barbados, Bangladesh and Ghana) and footwear. Consumer durables that were the subject of particular studies include household utensils, bicycles and furniture. The study on urban transport is rather different from the others, focusing on a basic service. All the goods and services studied have the common feature that they are required to a greater or lesser degree by all income groups and might normally be considered as worthy of expenditure even by the poor.

Such a variety of goods, services and countries may make interesting reading, but does not lead directly to general conclusions. These might have been more obvious had eight studies been carried out on the same product, although the significance of any conclusion based on such a series of studies would have been particularly limited. Nevertheless, we attempt in section 10.1 to find some common threads in the arguments presented in the various case-studies, linking their findings to the hypotheses set out in Chapter 1.

10.1 Conclusions from the Case-studies

In Chapter 1 appropriate products were defined as those that can satisfy the basic needs of low-income consumers in an efficient manner and at a low price. The efficiency of basic-needs satisfaction depends directly on the quality of its product, its intrinsic characteristics and the efficiency of consumption: and indirectly on the generation of employment oppor-tunities, and thus on incomes, for the under- and unemployed poor. Indeed, it is this last aspect, as explained in the Introduction, which led us to consider the questions of appropriate products and appropriate technology together.

The main themes in the case-studies are summarised below:

1. The first theme concerns the question of basic-needs characteristics and income classes. Chapter 1 developed the basic proposition that low-income consumers buy products and services that incorporate relatively more basic-needs characteristics than products and services purchased by high-income classes. The first problem faced by most of the authors of case-studies was to operationalise and measure the basic-needs characteristics or 'quality' incorporated in the various products. This is very difficult in the case of consumer durables such as footwear, metal utensils and furniture. It can be argued that the need satisfied by furniture, for example, is comfort and that its quality is related to the texture and finish of the wood, design, padding and upholstery, but how can such attributes be measured? In such circumstances it is necessary to resort to indirect measures of quality, such as the materials used (see footwear and metal utensils), or enterprise class (see furniture) or the technology used (see the study on soap by Mubin and Forsyth). In the study on food (weaning food), it might have been possible to measure

the number of proteins and calories per unit weight but this would have required a more extensive technical analysis than was in fact possible. The research on soaps by James sought to quantify the quality content by asking questions about consumer perceptions of eleven product characteristics including cheapness, whitening ability, attractiveness of scent, stain-removing ability, hand care, ease of lather, etc. Similarly, respondents to Fong's questionnaire to bicycle-owning households furnished judgements about seventeen possible attributes including price, maintenance cost, reliability, ease of getting spare parts, etc. James used a multiscaling model in order to compare the four products in the Barbados soap study and Fong used principal component analysis in order to group the characteristics into six underlying dimensions.

Most of the case-studies suggested that low-income consumers tend to buy the products most clearly embodying basic-needs characteristics. An exception is the study of footwear in Ghana, where low-income consumers prefer prestige and status and high-income households prefer durability. The reason for this is that people want to wear elegant shoes at social and ceremonial occasions. The study on metal utensils also shows that the required quality of a product is a function of its end-use. In this case, it is mainly cooking utensils that can be considered as satisfying basic needs; this is not so for serving utensils and other accessories.

Finally, in the study on soap in Barbados, James tries to determine the extent to which the consumers' perception of alternative products is shaped by publicity and to compare the 'subjective' versus the 'objective' perception of product characteristics. He finds that the brand 'Breeze' is perceived as fairly gentle, 'Drive' as a powerful stain- and soil-removing product, and 'Blue Soap' as cheap. Laboratory examination failed to confirm the strength of 'Drive' and the mildness of 'Breeze'; however, although the stain- and soil-removing ability of 'Drive' was no better than that of other detergents it did wash whiter. The discrepancy in perception must be due to advertising. The consumers accurately perceived the low price of 'Blue Soap'. Finally, 'Tide' is similar in quality to 'Breeze' and 'Drive' but is less popular, probably because of the lack of advertising and its slightly higher price. James also found that poor households use detergents less efficiently than the high-income consumers, often not reading the instructions on the packet, and using them for non-laundering purposes such as dish-washing and household cleaning.

2. The second theme is the question of whether low-income con-

sumers are unable to satisfy their basic needs efficiently because certain (appropriate) products and services are not available. Again, James documents the fact that until 1973 a cheap laundry soap (called 'Bomber') was on sale in Barbados but this was removed from the market to create room for imported products such as 'Sunlight' and 'Palm'. (However, according to the consumer survey, some 40–50 per cent of the poor still preferred 'Bomber' in 1979). On the basis of a so-called 'nested logit model', Thobani estimates that there is a frustrated demand of about 20 per cent for minibuses and of about 10 per cent for buses in Karachi. This situation arises because buses and minibuses cannot reach certain areas because of their physical inaccessibility or because current routes do not pass through those areas. Mubin and Forsyth encountered the reverse situation: they observed in Bangladesh that it was the greater availability that induced at least 35 per cent of the rural households to buy the 'labour-intensive' soap instead of the 'capital-intensive' variety.

3. Most of the eight case-studies take the role of marketing into consideration. Several suggest strongly that it is not only production technology that accounts for the differing performance of the modern and informal sectors, but also the different approaches to marketing. From a theoretical point of view, one could even claim that the concept of technology should not be limited to the production of goods and services but also to their marketing and distribution. This difference is brought out most clearly in the study by Aryee, in which it is noted that ex-factory prices of footwear of the same quality were about the same in small and large firms in Ghana, but that retail prices in the modern sector were about 50 per cent higher than in the informal sector. Because the informal sector tends to sell its products directly from the workshop, hardly any marketing costs are involved, whereas the modern sector sells its products through established wholesale and retail outlets markets, thus increasing the final retail price. Fong observes that the prices of bicycles are similar throughout Malaysia and therefore concludes that the distribution system works reasonably efficiently. Mubin and Forsyth stress that the labour-intensive sector in Bangladesh does not have a properly organised marketing system, often relying on hawkers and street vendors as retail outlets. In fact, only three of the labour-intensive enterprises in their survey pay for advertising. However, some have long-established connections with traders in distant rural areas. On the other hand, we already mentioned that for many rural customers it was not possible to buy capital-intensive products. In their analysis of the soap market in Ghana Mubin and Forsyth come to a similar

conclusion, but they find that on average people buy more capital-intensive washing soap than in Bangladesh and that the difference between urban and rural consumption patterns is less pronounced. Papola and Sinha find that metal utensils in rural India are about 20 per cent more expensive than in urban areas. This is because urban retailers buy either from the manufacturer or from the wholesaler, while the rural retailer buys either from the wholesaler or from the urban retailer. Moreover, as is also noted in several other case-studies, the rural consumer has to pay high transport and marketing costs. In Barbados, for example, relatively cheaper varieties of soap are not generally available in the small retail outlets, especially those in rural areas, because it is costly to distribute and sell small quantities to retailers in isolated rural localities. The marketing strategy does not only concern the physical distribution of the products and their price but it is also important to look at the quantities in which they are sold. It is well known that low-income consumers who receive their incomes irregularly or at daily or weekly intervals cannot afford to buy large quantities. James further reports that about 50–60 per cent of the urban and rural poor and about 40 per cent of the urban middle-income consumers in Barbados consider 16-ounce soap packages too large. According to James this is another reason for the inefficiency in soap purchasing.

4. A fourth theme concerns the question of whether low-income consumers buy appropriate products from the informal or small-scale sector. Most of the case-studies do indeed find that low-income groups often purchase goods produced in the informal sector. True, these products are not always the most appropriate, but they are usually low-priced and include more basic-needs characteristics than the corresponding modern-sector product. However, there are exceptions to this rule. The study on passenger transport in Karachi, for example, shows that the appropriate mode of transportation for low-income classes is the bus or the minibus, both of which can be considered part of the modern sector. The study on weaning food in Ethiopia shows that the choice is not between buying from the informal and formal sector but one between preparing weaning food at home or buying (or receiving) it from the modern sector. Finally, as was mentioned before, soap is no longer produced by traditional techniques in Barbados so that low-income consumers also buy their soap from the modern sector.

At the other end of the income scale we find that high-income classes do not necessarily buy from the largest enterprises in the modern sector. House, for example, reports that the highest-income classes in Nairobi (expatriates and nationals) tend to buy furniture from middle-sized enterprises (50–99 employees) which produce imitation 'antique' and well-upholstered furniture. However, most of the other studies find that high-income consumers purchase either from large domestic enterprises or from abroad.

5. A fifth theme concerns the question of whether appropriate products are produced with appropriate technology. In Chapter 1 appropriate technology was defined, albeit with some qualifications, as 'a set of techniques which makes optimum use of available resources in a given environment'. The satisfaction of basic needs within one generation was chosen as the criterion for optimality. At this stage, it should be remembered that technology is to be understood in the limited sense of production technology, and not in the wider sense embracing marketing. Two aspects of technology deserve further discussion here: the labour intensity of the technology and the kind of inputs it uses.

The fact that appropriate products are often produced by appropriate technology means at the same time that their production is relatively labour-intensive. We also found that the informal and small-scale sectors rarely produce goods for middle- and high-income consumers whereas large, capital-intensive enterprises often produce goods for purchase by low-income classes.

It may therefore be useful at this point to note the various technologies used in the traditional and modern sectors in some of the case-studies. For example, Fong shows that the automated operations of cutting, bending, pressing and threading employed by the Raleigh factory produce better quality than the equivalent manual operations in the Far East Metal Works. House finds that in the informal furniture sector sawing, smoothing and sanding operations are mainly carried out by hand, while they are carried out by machine in the largest enterprises. However, the middle-sized furniture enterprises apply the most capital-intensive production techniques, employing electrically driven tools. As was mentioned before, it is this class of enterprise that produces for the high-income market and which pays the highest salaries. Mubin and Forsyth also find that capital-intensive factories have a clear technical advantage over their labour-intensive rivals in virtue of their facilities for bleaching and cleansing which enable the removal of impurities from raw materials. Moreover, the labour-intensive sector is not able to produce toilet soap because the capital-intensive sector works with a special 'milling' process which permits moist extraction and the addition of scent and perfumes.

Papola and Sinha find that the rich man's metal for household utensils is stainless steel which is the most capital-intensive possible raw material. On the other hand, the utensils made of the poor man's metals (aluminium and iron) are produced by a more capital-intensive process than those made of brass and *phool*.

Not unexpectedly, therefore, one may conclude that large enterprises generally apply a relatively capital-intensive technology, that their products are of a better quality and that they pay relatively high wages. One can also assume (as was analysed in particular detail by Mubin and Forsyth) that the private profitability of capital-intensive technology is higher than that of labour-intensive technology. Finally, there was some supporting evidence for Stewart's thesis[1] that product characteristics are intimately linked to technology. However, it is our impression that this link is not a deterministic one but the result of the socio-economic environment prevailing in most developing countries. When the policy implications of the case-studies are discussed in the second part of this chapter, an attempt will be made to indicate how the apparent link between the choice of products and the choice of technology could be loosened.

Appropriate technology is defined not only in terms of the associated labour input but also by its use of inputs. The case-studies generally find that the large enterprises in the modern sector tend to import a larger part of their investment and intermediate goods than medium- or small-scale enterprises. For example, Fong reports that Raleigh in Malaysia imports $2.1 per dollar of internal value added, as against $0.6 for the non-Raleigh firms. Both the studies on soap confirm this, and the study on footwear establishes that in Ghana a substantial proportion of the raw materials, often in the form of unit soles, are imported. As Aryee points out, this dependence on imports has made the footwear industry vulnerable: indeed it was operated at a low rate of capacity utilisation since 1975 because of the shortage of foreign exchange. However, there are exceptions to this rule. Landgren, for example, shows that the large Faffa weaning-food factory obtains its raw materials in Ethiopia although the equipment was almost all imported (donated).

Mubin and Forsyth make an interesting but contrary point. They feel that the labour-intensive sector can only compete efficiently with the capital-intensive sector if the former has the same access to imported high-quality fats. An important reason for the higher quality of capital-intensive laundry soap is that it includes superior-quality fats which increase the stability of the lather and the firmness of the soap and reduce the dissolving rate.

6. Estimates of the direct employment impact of a (static) redistribution of income in consequence of a change in the pattern of demand

for substitute products were attempted in several of the case-studies. Mubin and Forsyth conclude that income redistribution is likely to have a small effect on employment through a shift of demand from capital-intensive to labour-intensive soap. However, they do not estimate the indirect employment effects, which are likely to be positive, because the labour-intensive sector tends to use more domestically produced inputs. Since the labour-intensive sector does not have any spare capacity, new investments in this sector would be needed, while capacity underutilisation would become more marked in the capital-intensive sector. Moreover, much also depends on which production units within the labour-intensive sector would benefit from increased demand: some inefficient plants use twenty times more labour per unit of output than the efficient ones. However, the entire argument is somewhat heroic because it would be difficult to change the urban consumption pattern back to labour-intensive soap, the advertising and distribution system being well organised in urban areas. Another very important point made by Mubin and Forsyth is that the potential for employment generation is far greater in the context of increasing demand consequent on the growth of population in the period up to 1990. That is to say, the employment impact of income redistribution is small compared with the impact of increasing demand over time.

The effects of a changing distribution of income on employment in furniture-making were analysed in a rather different way in the study by House on Kenya. There, the issue of a static redistribution of income was avoided. Instead, several estimates of employment generation in furniture-making were developed using different assumptions about the way in which future increments to national income would be distributed among the different income groups in the country. In particular, it was shown that if all the increase in national income in future years is attributed to the poorest 80 per cent, the rate of growth of employment would be 6.1 per cent in contrast to 4.6 per cent if income increases were distributed according to the 1976 pattern.

The distinction between the analysis of a static income redistribution and a dynamic one is brought out clearly in the study on metal utensils by Papola and Sinha. There it is shown, in contradiction to a principal hypothesis in this volume, that a static redistribution of income in favour of the poor would lead to reduced production and employment in the metal utensils sector, because low-income groups tend to spend a smaller proportion of their income on utensils than high-income groups. Moreover, the latter purchase more expensive utensils, so that the actual demand in terms of the number of units is less than if the same amount of income was spent by low-income groups. On the other hand, increasing incomes over time among all income groups will inevitably lead to increased employment, although the proportion of utensils with non-basic-need attributes would also tend to increase.

10.2 Some Suggestions for Future Research

It may now be useful to summarise the various suggestions for future research made above. First of all, more research could be carried out on the perception of low-income consumers. Is it correct that low-income consumers (both in urban and rural areas) are more likely to be influenced by promotional activities (such as advertising) than high-income consumers? James examined this question of the consumption inefficiency of the poor in the case of soap in Barbados. However, more studies on other products and other countries are needed in order to be able to generalise his findings. James also suggests that more education may be the best remedy for consumption inefficiency. This hypothesis also merits further examination in other countries.

Secondly, there is a need to make the concepts of quality and basic-needs product characteristics more operational. As mentioned before, this need is most evident in respect of consumer durables. The accurate definition and measurement of these characteristics would enable governments in developing countries to define appropriate quality standards.

Thirdly, little is known about the marketing techniques of informal and small-scale enterprises. It was established in most of the case-studies that these enterprises mainly sell their products to low-income consumers. However, by studying their sales techniques, one might be able to determine how they can enlarge their markets.

Fourthly, there is room for more product case-studies that include an analysis of the choice of products and of technology. Such analyses may be very useful to governments interested in formulating the sort of comprehensive product policy described in the next section.

Fifthly, these product studies may be complemented by engineering studies on particular products. The latter would provide designs of appropriate products either adapted from existing products or newly invented. Such studies would also provide a detailed analysis of the technology to be employed for producing these products.

Finally, it would be useful to carry out case-studies on particular policies to stimulate the production and consumption of appropriate

products. Examples of such policies are consumer education programmes; codes of advertising; tariff and import policies; tax and subsidy
policies; investment and other production incentives; price policies;
marketing and distribution policies, etc.

10.3 Policy Implications

Not all the case-studies lead directly to conclusions on the policy
implications of the research findings. However, it is clearly essential to
examine these implications because they may enable governments,
producers and consumers to contribute to greater consumption efficiency and the creation of additional employment opportunities. The first
part of this section will concentrate on policies leading to a better choice
of products while the second part will concentrate on a better choice of
technology.

Choice of products

The first set of policies to be discussed is *consumer perception and
consumer efficiency.* A considerable body of evidence suggests that
consumer perceptions of various products and their subsequent purchases are shaped by advertising and experience. As James points out,
education is likely to be the most effective means of improving the
purchasing skills of low-income groups. He found that the cause of some
consumption inefficiency lies in the attitudes of the poor to the
consumption process in general. At present, most educational systems
do not pay attention to consumer education. However, it is difficult to
specify in detail the content of consumer education programmes for
school children of different ages. Nevertheless, since most children from
the poorest households do not reach secondary level, it is important that
this type of education be introduced in primary schools. Non-formal
adult education programmes are equally desirable to change attitudes
and improve consumer skills. Such programmes may be organised in the
context of adult education programmes or by organisations such as co-
operatives, youth movements, women's groups and consumer
organisations.

A second condition for efficient consumption is *a reasonable balance
between various information sources.* However, because of advertising
the consumer typically receives a good deal of information on products
and services from the modern sector but very little on those from the

traditional sector. It is the responsibility of producers to ensure, first, that the information provided by advertising is factually correct; secondly, that it encourages the consumer to buy a product for the right reason. One useful way of achieving this main aim is a code for advertisers drawn up jointly by producers, governments and consumer organisations. Such a code may also include provisions to persuade its application as widely as possible.

A further step would be for a government to formulate a product-information policy, perhaps in consultation with producer and consumer organisations. Such a policy could include television and/or radio programmes, which are probably the most effective means of communicating information about product characteristics. To be effective, this information should be comparative in nature, product by product. There is little point in merely conveying to consumers the characteristics embodied in a large number of different product brands. The policy could also include the limitation of the advertising of less appropriate products and, conversely, increased advertising of more appropriate products.

A third important policy issue relates to *product standards* and the use of *trade marks*. Some goods, for example fresh food and simple household equipment, lend themselves more easily to efficient choices because their characteristics can be assessed accurately prior to purchase. On the other hand, there are others, including pharmaceuticals and consumer durables for example, the attributes of which can be discovered only with difficulty, even after use. The case for product standards is strongest where there are acute difficulties in establishing the quality of the product and/or where the consequences of inefficient purchase decisions by consumers are serious. As James and Stewart[2] have observed, in many countries government standards/specifications for products are either totally absent or are imitations of the corresponding standards in developed countries, leading to overly sophisticated high-income products. It is desirable that government standards be designed so that basic-needs-type characteristics are embodied in the relevant products.

Trade marks[3] are important in product differentiation, 'a strategy by which a firm aims one type of product at the total market and attempts to establish in the customer's mind the superiority and preferability of this product relative to competing brands'.[4] Thus the main function of a trade mark is the creation of goodwill among consumers, and a secondary function is to identify products and indirectly their quality. The principal beneficiary is therefore the owner of the mark, often a foreign company. Such a company is often rewarded handsomely for its advertising efforts, the price being paid by the consumer. Various options are open to reduce the impact of trade marks and to increase its information function. In industries where there is a strong brand proliferation, such as the pharmaceutical, food and tobacco industries, one could consider the withdrawal of trade mark protection. In the case of pharmaceuticals for example, one could switch from brand names to generic names, a policy that has been applied in Sri Lanka and is under consideration in India.[5] It may also be possible to charge higher initial and renewal registration fees in order to discourage the use of trade marks. In order to improve their quality-control function for consumer goods, product marks could be instituted, the quality of which would be registered with the trade mark.

A second set of policies relates to the *price and availability of products*. As noted by James and others, poor consumers often cannot afford to buy in large quantities. If products are marketed in small quantities, the implicit price per unit is frequently higher because of higher marketing costs. This occurs in the developed countries[6] and is also likely to occur in the developing countries. It may therefore be desirable for governments to encourage enterprises, by tax and other incentives, to market their products in small quantities to the extent appropriate. More generally, a government may adopt a price policy towards appropriate products and subsidise some of them. For example, many countries seek to control the prices of basic foodstuffs, although such policies are frequently effective only in the short run. If price controls are maintained over a long period and if the underlying supply and demand factors are left unchanged, price controls are frequently counter-productive. On the other hand, if the public sector is an important supplier of a service – urban transport for example – the impact of a pricing policy may be considerable. For example, in Karachi Thobani estimates that decreasing the waiting time for buses (through direct supply) and minibuses (through licences) by 10 per cent will lead to an increase of 40 per cent in bus demand and 70 per cent in minibus demand. He also establishes that a change in the fares of buses and minibuses has a smaller impact on demand than the same percentage change in waiting time. It may also be desirable for governments to formulate a comprehensive product policy. In some cases it may make good sense for governments to subsidise products or services which incorporate mainly basic-needs characteristics, and this issue is further discussed below.

Most products available on the world market were originally designed for consumers in high-income developed countries. They satisfy a great variety of needs, many of which are not basic in developing countries. Thus, there is a good case for the developing countries to reduce the number and availability of those products on their domestic market. On the other hand, it will usually be unrealistic to attempt to ban 'high-income' products altogether: given the considerable inequality of incomes in many developing countries, the effective demand for such products is substantial.

The first step in a consistent product policy[7] would be to define which products or product ranges are required. Having decided that the country needs the product, the second question is whether it should be imported or produced domestically. If the product includes many high-income characteristics and it is decided to import it, then the government may impose a high import tax. This occurs frequently in the case of products such as cars, televisions and similar obviously luxury consumption items. If it is decided that the product is to be produced domestically, the next question is what type of enterprise should produce it. The choice is then between small or large domestic enterprises, or international companies. In the last case, one has to determine the optimum way of assembling the package of capital, know-how, management and market access. If small or large domestic enterprises are preferred, the technology issue has to be examined, that is, to what extent do these enterprises need scarce capital resources and to what extent do they create employment (directly or indirectly) and do they use mainly local inputs? These issues will be discussed in detail in the second part of this chapter, where we deal with policies regarding the choice of technology. However, there are also certain intermediate solutions. For example, in India the production of simple radios has been allocated to the small-scale sector, but more sophisticated ones are produced by the subsidiaries of certain multinational enterprises.

Choice of technology

Most of the case-studies reveal that the technology embodied in the production process of informal and small-scale enterprises is more labour-intensive and uses more local inputs. However, the disadvantage of labour-intensive techniques is that productivity is relatively low, and wages remain low in consequence. Moreover, the income elasticity for products produced by this sector is low. Several of the case-studies therefore propose policies to increase the productivity of the domestic and small-scale sector and improve its marketing capabilities. A second set of policies also considered in this section concerns research and development on new products appropriate to developing countries.

A high priority for most developing countries is the *strengthening of the household enterprise and small-scale industry sector*. Although it is not appropriate here to develop a comprehensive policy towards technology and the informal sector,[8] it may be useful to emphasise several related issues discussed in the case-studies. One of these is the role of marketing. For example, Fong notes that the existing industrial incentives in Malaysia favour either very investment-intensive or very labour-intensive technology but discriminate against intermediate technology. He also recommends a balanced tariff policy, for two main reasons. First, such a policy may to some extent protect labour-intensive production from international competition. Secondly, it may encourage increases in productivity which would eventually make these products more competitive on the international market. Mubin and Forsyth conclude that small-scale enterprises would be able to compete more effectively with the large-scale enterprises if they had the same import facilities. In present circumstances, the quality of laundry soap produced by small-scale enterprises is lower than that produced by large enterprises since the latter can use superior-quality fats as input materials. Finally, House makes several recommendations relating to the improvement of the performance of the informal sector. His proposals concern, for example, the provision of infrastructural services (including, *inter alia*, electricity, transport facilities, roads, permanent structures and/or mini-estates) and credit facilities both for the purchase of equipment and for working capital.

One of the main findings of this book is the difference in performance between the informal and the formal sector. This is a question of *marketing strategy* as well as production technology. Whereas the formal sector spends considerable sums on advertising and physical distribution for a wide and open market, the informal sector tends to be able to sell its products only to customers who present themselves at the workshop, or through hawkers or street vendors. More research is certainly needed on this issue, and the suggestions that follow are offered somewhat tentatively.

The main difficulties in marketing the products of informal and small-scale enterprises are the following:

- there are many scattered enterprises which produce in small quantities;
- the products are uneven in quality and cannot be recognised as such;
- the rate of production is often irregular;
- the customers (particularly in rural areas) are difficult to reach.

From a marketing point of view the advantages of these enterprises are that the prices of their products are low, that they are flexible, that there is an immediate group of customers and that products can be made according to specifications established by individual customers. In consequence, there is a dilemma in the sense that if informal enterprises were to produce products of the same quality with a brand name that could be recognised, they would lose some of the private and social advantages of their informal sector status.

At this point, it may be useful to make a distinction between rural areas and small towns on the one hand and (large) cities on the other. It seems appropriate for small enterprises in rural areas and provincial towns to produce for customers in the immediate surroundings. These customers can be served either directly at the workshop or in local market-places. However, it seems that in cities small enterprises could benefit from the existence of a large market in an especially obvious manner. Such benefits could only be realised if there is some form of organisation. Several small enterprises could, as is suggested by House with regard to Kenya, establish themselves on mini-estates. By forming a co-operative or otherwise organising themselves they could agree on the production of products to a certain quality standard and on co-ordinated sales efforts both to customers on the doorstep and to wholesale and retail traders. Governments can also encourage informal-sector production directly. House suggests, for example, that the government should endeavour to purchase goods from the informal sector as well as modern, large-scale industries in the formal sector.

A second group of policies concerns *research and development on new products*. As James and Stewart[9] observe, 'there is a systematic tendency for product development in high-income countries to be inappropriate for developing countries, leading to products containing excessive characteristics. The development of appropriate products with low-income characteristics (including divisibility) and appropriate to the environment of poor countries would enable such countries to benefit from the increased efficiency of modern products without being harmed by their high-income characteristics'. Some of the case-studies in this volume provide specific examples of the need for new products appropriate for low-income consumers. For example, Landgren-Gudina mentions the home-made weaning food prepared according to recipes from the Ethiopian Nutrition Institute. Mubin and Forsyth point out that small enterprises in Bangladesh could produce better-quality laundry soap if they were allowed to import high-quality fats. James notes that in Barbados detergents are marketed in packets that are too large for low-income consumers. Finally, Fong shows that the largest domestic bicycle manufacturer in Malaysia would produce better-quality products and would be better able to compete with the Raleigh manufacturer if it mechanised its cutting, bending, pressing and threading operations.

In order to identify new appropriate products and services, collaboration between the marketing economists and production engineers is required. Marketing studies may be useful in respect of products or product ranges that are bought both by low- and high-income consumers. By comparing the basic-needs characteristics included in the products, it is possible to determine whether the products presently available efficiently satisfy the needs of low-income consumers. The engineer has a role to play in examining whether the quality of existing products can be improved or adapted, or whether it is possible to design products with a new combination of basic-needs characteristics. Following advice from an economist, the engineer may be able to determine the most appropriate technology for producing these products. On the basis of such designs, the marketing economist could then investigate the effective demand for such products. If an effective demand is not evident and if the products nevertheless increase the consumption efficiency of low-income classes, the government may consider subsidising its production.

It is difficult *a priori* to determine the type of organisation in which such a collaboration of engineers and economists described above should take place. There are several possible arrangements. For example, in some developing countries, governments already subsidise consumer organisations and the latter could in addition be encouraged to engage in product development. Another possibility is to subcontract these activities to national technology centres and technical institutes attached to universities. Finally, it may be useful for research institutes in the public sector to collaborate with enterprises from the private or public sector that are normally in close touch with the consumer, although not necessarily with the poorest among them.

Notes

1. F. Stewart, *Technology and Underdevelopment* (London: Macmillan, 1977).
2. J. James and F. Stewart, 'New Products: A Discussion of the Welfare Effects of the Introduction of New Products in Developing Countries', *Oxford Economic Papers*, vol. 33, no. 1 (Mar 1981) pp. 81–107.

3. A comprehensive analysis of trade marks is given in UNCTAD, *The Impact of Trade Marks on the Development Process of Developing Countries* (mimeo.) (Geneva, June 1977, TD/B/C.6/AC.3/3).
4. W. R. Smith, 'Product Differentiation and Market Segmentation as Alternative Marketing Strategies', *Journal of Marketing* (July 1956) pp. 3–4, quoted in W. M. Pride and O. C. Ferrall, *Marketing Basic Concepts and Decisions* (Boston: Houghton Miffin, 1977) p. 65.
5. UNCTAD, *The Impact of Trade Marks*, p. 89.
6. D. Caplowitz, *The Poor Pay More* (New York: Free Press, 1963); and H. Kunreuther, 'Why the Poor May Pay More for Food: Theoretical and Empirical Evidence', *Journal of Business*, vol. 46, no. 3 (July 1973) pp. 368–83.
7. Many of the ideas on product policy mentioned here are drawn from P. Streeten, 'Issues for Transnational Corporations in World Development', *The CTC Reporter*, vol. 1, no. 10 (New York: Centre on Transnational Corporations, Spring 1981) pp. 19–22.
8. See, for example, S. V. Sethuraman (ed.), *The Urban Informal Sector in Developing Countries* (Geneva: ILO, 1981); and A. S. Bhalla (ed.), *Technology and Employment in Industry*, 2nd edn (Geneva: ILO, 1980).
9. J. James and F. Stewart, 'New Products: A Discussion of the Welfare Effects of the Introduction of New Products in Developing Countries', *Oxford Economic Papers*, vol. 33, no. 1 (Mar 1981) p. 104.

Index